GW00578214

THE CHINESE SPACE PROGRAMME

From Conception to Future Capabilities

WILEY-PRAXIS SERIES IN SPACE SCIENCE AND TECHNOLOGY

***Series Editor:* John Mason, B.Sc., Ph.D.**

This series reflects the significant advances being made in space science and technology, including developments in astronautics and space life sciences. It provides a forum for the publication of new ideas and results of current research in areas such as spacecraft materials, propulsion systems, space automation and robotics, spacecraft communications, mission planning and management, and satellite data processing and archiving.

Aspects of space policy and space industrialization, including the commercial, legal and political ramifications of such activities, and the physiological, sociological and psychological problems of living and working in space, and spaceflight risk management are also addressed.

These books are written for professional space scientists, technologists, physicists and materials scientists, aeronautical and astronautical engineers, and life scientists, together with managers, policy makers and those involved in the space business. They are also of value to postgraduate and undergraduate students of space science and technology, and those on space-related courses (including psychology, physiology, medicine and sociology) and areas of the social and behavioural sciences.

For further details of the books listed below and ordering information, why not visit the Praxis Web Site at http://www.praxis-publishing.co.uk

METALLURGICAL ASSESSMENT OF SPACECRAFT PARTS, MATERIALS AND PROCESSES
Barrie D. Dunn, Head of Metallic Materials and Processes Section, ESA-ESTEC, Noordwijk, The Netherlands

SATELLITE CONTROL: A Comprehensive Approach
John T. Garner, Aerospace Consultant, formerly Principal Ground Support Engineer, Communications Satellite Programmes, ESA-ESTEC, Noordwijk, The Netherlands

SOLAR POWER SATELLITES: A Space Energy System for Earth
Peter E. Glaser, Vice President (retired), Arthur D. Little, Inc., USA, Frank P. Davidson, Coordinator, Macro-Engineering Research Group, Massachusetts Institute of Technology, USA, Katinka I. Csigi, Principal Consultant, ERIC International, USA

THE MIR SPACE STATION: A Precursor to Space Colonization
David M. Harland, formerly Visiting Professor, University of Strathclyde, UK

LIVING AND WORKING IN SPACE: Human Behavior, Culture and Organization, Second edition
Philip Robert Harris, Executive Editor, *Space Governance* Journal; Vice President, United Societies in Space, Inc., USA

THE CHINESE SPACE PROGRAMME: From Conception to Future Capabilities
Brian Harvey, M.A., H.D.E., F.B.I.S.

THE NEW RUSSIAN SPACE PROGRAMME: From Competition to Collaboration
Brian Harvey, M.A., H.D.E., F.B.I.S.

Forthcoming titles in the series are listed at the back of the book.

THE CHINESE SPACE PROGRAMME

From Conception to Future Capabilities

Brian Harvey, M.A., H.D.E., F.B.I.S.

JOHN WILEY & SONS
Chichester • New York • Weinheim • Brisbane • Singapore • Toronto

Published in association with
PRAXIS PUBLISHING
Chichester

The White House,
Eastergate, Chichester,
West Sussex, PO20 6UR, England

Published in 1998 by
John Wiley & Sons Ltd
in association with Praxis Publishing Ltd

Wiley Editorial Offices

John Wiley & Sons Ltd, Baffins Lane,
Chichester, West Sussex, PO19 1UD, England

John Wiley & Sons, Inc., 605 Third Avenue,
New York, NY 10158-0012, USA

Wiley-VCH Verlag GmbH, Pappelallee 3,
D-69469 Weinheim, Germany

Jacaranda Wiley Ltd, G.P.O. 33 Park Road, Milton,
Queensland 4001, Australia

John Wiley & Sons (Asia) Pte Ltd, 2 Clementi Loop #02-01,
Jin Xing Distripark, Singapore 12981

John Wiley & Sons (Canada) Ltd, 22 Worcester Road,
Rexdale, Ontario, M9W 1L1, Canada

Library of Congress Cataloguing-in-Publication Data
Harvey, Brian, 1953–
The Chinese space programme : from conception to future capabilities / Brian Harvey.
p. cm. -- (Wiley-Praxis series in space science and technology)
"published in association with Praxis Publishing, Chichester."
Includes bibliographical references and index.
ISBN 0-471-97588-5 (alk. paper)
1. Astronautics–China I. Title. II Series.
TL789.8.C55H37 1998
387.8'0951—dc21 97-20317
CIP

A catalogue record for this book is available from the British Library

ISBN 0-471-97588-5

Printed and Bound in Great Britain by MPG Books Ltd, Bodmin

To my father and mother,
who encouraged my interest in spaceflight
during my childhood

Table of contents

Author's preface

On 24 April 1970, China launched its first Earth satellite, Dong Fang Hong ('the East is Red') number 1, astounding both its Pacific neighbours and western countries who wondered how a country in the midst of political turmoil and poverty could achieve such a scientific breakthrough.

It is not often realised that the Chinese developed their modern space programme at about the same time as the United States and the Soviet Union began their big push into space. The Chinese space programme officially began on 8 October 1956, a year before the first Earth satellite was even launched. On that day, the Chinese political leadership decreed the foundation of the Fifth Research Academy to spearhead China's space effort, and at once requisitioned two abandoned sanatoria in Beijing to be its first laboratories. Had it not been for political upheaval, in the form of the great leap forward and the subsequent Cultural Revolution, the Chinese might have achieved much more, much sooner.

China's first satellite in orbit was the biggest of all the space powers. China was the third space power to recover its own satellites, put animals into orbit and develop hydrogen-fuelled upper stages. Now, over 30 years later, China has launched more than 50 Earth-orbiting satellites. It holds a solid share of the world launcher market, and is preparing to be the third nation to launch its own cosmonauts, which will take place in 1999 to mark the 50th anniversary of the Chinese revolution.

The Chinese space programme has been sometimes called the last secret space programme. There is some truth in this, for China reminds one of what the Soviet space programme used to be like before glasnost opened up the hidden pages of its history. Some aspects of the Chinese space programme are now very open and there is an abundance of technical information. But in other ways it is still secretive, information is difficult to obtain and it too has its blank pages to be filled in. The historian particularly misses the absence of memoirs of the leading personalities of the programme.

Although the Chinese have confirmed some launch failures from the early days of their programme, there are many rumours of early launch failures which have been neither explained nor convincingly refuted. An entire series of satellites, the Ji Shu Shiyan Weixing, flown in the course of 1975–76, has never been explained or declassified. We know virtually nothing of the Shi Jian 2A and 2B satellites, launched in 1981, nor of the large and enigmatic KF-1 payload carried aloft by the Long March 3A with Shi Jian 4 in 1994.

Technical aspects aside, we know little of the often tumultuous interaction between the leaders of the programme and the political leadership of China. Although many of the designers of the Chinese space programme are known by name, and their biographical details are available, their characters and personalities remain inscrutable to this day. One must patiently put together all the different elements, trying to make a cohesive whole, but always be aware of the many pieces of the jigsaw which we still know too little about.

A complicating factor – one familiar to students of the Soviet and Russian space programme – is the use of designators for different satellites. In the west, the first Chinese satellites were designated China 1, 2, 3 and so on, but other terms were also used: PRC-1 (People's Republic of China 1), PRC-2, PRC-3; also Mao-2, Mao-3, and so on. At the time, the Chinese did not give designated titles to their satellites; they simply referred to the dates of launch. Eventually, the Chinese themselves introduced a set of designators, and these have been applied retrospectively. To complicate matters, the Chinese have also revised their own designators since, and have done so more than once. Not only satellites, but many Chinese space-related facilities have changed their names or titles. An additional problem is that the names of many institutes are translated slightly differently in different publications, creating the possibility of some confusion.

The Chinese space programme – from conception to future capabilities is the history of the Chinese space programme. Chapter 1 outlines it origins. Chapter 2 tells of how China put its first satellite into orbit. There was a significant expansion of the Chinese space programme in the decade which followed, marked by the introduction of recoverable spacecraft (chapter 3). Chapter 4 recounts how the Chinese introduced a powerful new rocket, the Long March 3, lofted communications and weather satellites and developed other new space activities in the 1980s and 1990s. In chapter 5, *Making it all possible*, there is a review of the Chinese space industry, facilities, organisation and infrastructure before the book goes on to examine China's rocket launchers, engines and launch sites (chapter 6). Finally, chapter 7 looks forward to China's future space capabilities, including manned flight.

Brian Harvey
Dublin, February 1998

Acknowledgements

In writing this book, I am most grateful to Rex Hall and Phil Clark, both of whom provided a range of useful information and shared generously of their information and files on the Chinese space programme. I wish to acknowledge and thank Phil Clark of the Molniya Space Consultancy for permission to use a number of graphics and diagrams taken originally from a number of Chinese publications. From China, I received valuable assistance from Dr Zhu Yilin of the Chinese Academy of Space Technology. Many other people, such as James Harford and Dominic Phelan, helped in the provision of information and advice. I wish to especially acknowledge Neil Da Costa, London, for the provision of photographs; the Great Wall Industry Corporation in Munich, Germany, for supplying an extensive range of photographs and illustrations, and Camera Press. I am grateful to them all.

List of illustrations, maps and tables

Chapter 5

Chapter 6

Chapter 7

Tables

1

Origins: the fiery dragon, 1000–1959

A history of space exploration in China is merited not least because the rocket – the word means 'firing arrow' in Chinese – was invented there. The ancient Chinese discovered the secrets of gunpowder in the seventh century. The rocket was invented by Feng Jishen in 970. Primitive rockets – which would now approximate to fireworks – were used by the Song dynasty to fight Xia in 1083. When the Japanese invaded China in 1275, Kublai Khan fired rockets to drive them away. The histories of the Ming dynasty (1368–1644) reported that over 39 types of rocket weapon were in use during the period, and recorded the use of a two-stage, four-tube rocket called by the navy the *fiery dragon* which was fired several thousand metres over water. Various combinations of rockets were built, some with poetic names such as *swarm of bees*, *magic flying crow* and *thunder cannon shocking heaven and flying into the sky to strike the enemy*, all devised in various fiendish ways to inflict terrible damage on China's enemies. The most terrifying must have been *magic firing box* (1377) in which 3,600 rockets were fired simultaneously. These ancient Chinese rockets possessed the fundamental elements of modern rockets: a combustion chamber, firing system, explosive fuels and feathers for guidance.

A sixteenth century inventor called Wan Hu designed a rocket-propelled chair on which he planned to ascend into heaven. He built an open cabin, to which he fitted 47 rockets underneath and above, and two kites to keep him aloft. Wan Hu disappeared in flame and smoke and was never seen again. A crater on the Moon is now named after him, so in one sense he made it to the heavens after all. This is the first recorded design of something approximating to a manned space rocket.

TSIEN HSUE-SHEN

Several names are irrevocably associated with the development of the world's great space programmes: in Russia, Tsiolkovsky, Korolev, Glushko, Chelomei; in the United States, Goddard, von Braun, Faget. The person who contributed most to the development of modern space exploration in China was Tsien Hsue-shen. Hardly known in the west, he was born in Hangzhou, Zhejiang in 1911, the only child of an education official. His name – which means 'study to be wise' in Chinese – reflected his father's ambitions. Much of what we know about Tsien comes from a pioneering biography by Iris Chang, *The thread of the silkworm*[1]. Tsien Hsue-shen attended a primary school for gifted children in Beijing. A model child with an outstanding school record, he subsequently entered the Beijing Normal University High School. At 18, he applied to Jiatong University in Shanghai to study railway engineering, coming third in the nation in the entrance exam. A serious, aloof, immaculately dressed and perfectly behaved student, Tsien was a man who liked to study and work on his own, his only outside interest being classical music (he played the violin). His time in Shanghai was punctuated by typhus (which he only narrowly survived), social turbulence and the beginnings of the war with Japan.

Graduating as top student, with 89 points out of 100, he chose to pursue aeronautical engineering, competing for a scholarship in the United States in 1935. There he started at Massachusetts Institute of Technology (MIT), staying only a year before moving to the California Institute of Technology (popularly known as CalTech) in Pasadena, where he studied under the great Austro–Hungarian mathematician Theodore von Kármán. Tsien graduated as doctor in 1939.

Here we can begin to trace Tsien's interest in rocketry. Five fellow students and associates invited him to join a group interested in what would now be called amateur rocketry: a gang of experimenters buying up spare parts, assembling them and letting them off in the

Tsien Hsue-shen.

nearby desert. He was in effect the mathematics adviser to the group, in 1937 writing his first work on rocketry, *The effect of angle of divergence of nozzle on the thrust of a rocket motor; ideal cycle of a rocket motor; ideal efficiency and ideal thrust; calculation of chamber temperature with disassociation.* Their first, often dangerous, experiments were presented to the Institute of Aeronautical Sciences and written about locally in the student press, where Tsien made some unguarded comments about the possibility of sending rockets 1,200 km into space. Rather like fellow rocketeers in Germany and the Soviet Union, their work soon became sponsored by the military, who saw the potential for rockets both to make aircraft fly faster and to fly as ballistic missiles. Military funding rose from $1,000 to $650,000 in five years. By 1942, after the United Sates entered the war, Tsien was working on small solid rocket motors to help aircraft get airborne; shortly afterwards he helped to draw up plans for a missile programme.

Tsien became Assistant Professor of Aeronautics in 1943. As well as doing research, he taught students, though many found his manner intimidating, intolerant, arrogant, over-precise and unsympathetic. He was one of the co-founders of the Jet Propulsion Laboratory (JPL), from where American unmanned exploration of the Moon, the nearby planets and the outer Solar System was to be subsequently guided. He was the first head of research analysis at JPL in 1944, and by the following year was working for Kármán in the Pentagon advising the United States military on how to harness the latest discoveries in aeronautics and rocketry for the post-war defence forces.

ON THE TRAIL OF THE GERMAN ROCKETEERS

In May 1945, having been given the rank of temporary colonel of the United States Air Force, Tsien arrived in Germany to survey the Nazi wartime achievements in rocketry, their rocket factories and secret test sites. On 5 May he met the leading German rocket engineer, Wernher von Braun, who had just surrendered to the Americans. Not long afterwards, the man who was to be his opposite number in the Soviet Union, chief designer Sergei Korolev, was scouring Germany on an identical mission.

Returning to JPL, Tsien published his wartime technical writing in a book called *Jet propulsion*. After a return to MIT in 1946–48 and a brief visit to China in 1947 (where he married), he became in 1950 the Robert Goddard Professor of Jet Propulsion at Cal-Tech. He gave a presentation to the American Rocket Society in which he outlined the concept of a transcontinental rocketliner able to fly 400 km above the Earth, its space-suited passengers floating in its cabin as they briefly enjoyed weightlessness. This presentation was later covered in *Popular science*, *Flight* and the *New York Times*. The following year, he predicted that astronauts would travel to the Moon within 30 years. Some of his rocketplane ideas inspired the United States Air Force to develop its space-plane project of the late 1950s, the Dynasoar, ultimately one of the ancestors of the space shuttle.

The following year, at the height of the McCarthy witch-hunt in the United States, Tsien was accused of being a communist. His security clearance was revoked, which pained him greatly. He was put in jail and then held under house arrest. For many months, the different factions of the American government battled over whether he should be released, put back in jail or deported. For Tsien, this was a period of great confusion and uncertainty, but he kept up his work on rocket guidance and on how computers could steer rockets during their ascent through the atmosphere.

RETURN TO CHINA

In a September 1955 agreement between the American and Chinese governments, Tsien and 93 fellow scientists returned to communist China in exchange for 76 American prisoners of war taken in Korea. Reentering China through Hong Kong (then a British colony) Tsien and his family were warmly greeted in Shenzhen by the Chinese Academy of Sciences and welcomed in a series of homecoming celebrations that culminated in Beijing, just restored as China's capital city.

It is no coincidence that the beginning of China's modern missile programme may be dated to 1956, the year after Tsien's return. He brought with him the most up-to-date theory of rocketry from the United States. However, he had to start virtually from scratch. China in the early 1950s had barely emerged from a long period of great turbulence and destruction – social unrest, the war with Japan, then the civil war and finally the communist revolution of 1949. Making cars and trucks represented the limit of China's industrial and technical capacity; there were no aircraft factories, test sites, wind tunnels or the type of facilities Tsien had begun to take for granted in California. However, reconstruction and modernisation had begun, the first five-year plan for socialist reconstruction having been adopted in 1953.

SCIENTIFIC CONSTRUCTION IN CHINA

In January 1956, Chairman Mao Zedong proposed the rapid development of science and technology in China so that the country could catch up quickly with the world's most advanced levels in economics and science. In response, the Supreme State Conference set up a Scientific Planning Commission under the leadership of prime minister Zhou Enlai. In February, Tsien presented a report to the Central Committee: *Opinion on establishing China's national defence aeronautics industry*. The commission consulted widely with scientists and experts, after which it drew up *Long-range planning essentials for scientific and technological development, 1956–67*, which defined 57 priority tasks which would ensure China's independence in rocket and jet technology in 12 years. These included atomic energy, rockets, jet technology and computers.

In April 1956, Zhou Enlai presided over a meeting of the Central Committee Military Commission which invited Tsien Hsue-shen to outline the potential of guided missiles and rockets. Within days, the government had appointed a State Aeronautics Industry Commission to develop the country's aviation and missile defences. Deputy premier Nie Rongzhen was made director and Tsien Hsue-shen was one of its members. On 10 May, Nie Rongzhen issued his first report: *Preliminary views on establishing China's missile research*. On 26 May, the Central Military Commission, chaired by Zhou Enlai, accepted the report and ordered the setting up of administrative machinery to get work under way.

FOUNDATION: 8 OCTOBER 1956

On 8 October 1956, the Central Committee of the Communist Party of China, presided over by Mao Zedong, established the Fifth Research Academy of the Ministry of National

Defence to develop the space effort. This is now officially marked in China as the birth day of the Chinese space programme. Within the academy, the first Rocket Research Institute was established under vice-premier Nie Rongzhen, with Tsien as its first director. The government took over two abandoned sanatoria and requisitioned 100 university graduates to begin work there. Nie Rongzhen reported to the Central Committee on its establishment later in the month.

Leaders of the Fifth Research Academy of the Ministry of National Defence, Beijing, 1956

Tsien Hsue-shen
Guo Yonghuai
Xu Guozhi

THE FIFTH ACADEMY

Even though it had such an enigmatic title, the existence of the Fifth Academy was also classified. This was a Russian trait too; the space industry there operated under the sonorous title of the Ministry of General Machine Building. Likewise, the body responsible for nuclear development was similarly classified (the Ninth Academy; in Russia it was the Ministry for Large Machine Building).

The Fifth Academy's first task was not to experiment, as its leaders might have liked, but to build up a basic expertise among staff who were keen but had not had the opportunity to reach a third level education. Tsien was joined by a number of his contemporaries, some of whom had studied while he was in college and many of whom had, like him, worked in the United States. Typical of these were fellow rocket experts Liang Shoupan, Tu Shoue and guidance expert Huang Weilu. The Fifth Academy set up ten research laboratories and two branch academies to deal with the many aspects of designing modern missiles and preparing them for flight.

The leadership of China soon realised that the institute would still need outside help if it was to progress at all. No matter how good Tsien's theoretical knowledge and that of his colleagues, China simply did not have the ability to develop even basic rocket hardware from scratch. The leadership therefore turned to China's communist ally, the Soviet Union.

CHINESE–SOVIET COOPERATION

Negotiations with the Soviet Union began in 1956. The Russians were then at a critical stage in the development of their own intercontinental ballistic missile and Earth satellite project[2]. In September 1956, Nie Rongzhen led a delegation to visit Moscow. The Soviet Union agreed to sell China two R-1 missiles, which were essentially Soviet replicas of the German A-4 wartime weapon better known as the V-2. These duly arrived in October 1956. However, they cannot have been much help to Tsien in his plans, for he had already seen A-4s during his visit to Germany eleven years earlier.

The Chinese considered that the R-1 did not represent the most up-to-date Soviet technology then available. Of course, they were right; the Russians had moved beyond the R-1 in 1949 and had now reached the R-5. In Dnepropetrovsk, the Mikhail Yangel Design Bureau (now known as NPO Yuzhnoye) had modified the R-1 and considerably improved it as the R-2. This missile had a range of 590 km.

China appealed to the Soviet Union to be more forthcoming, and Nie Rongzhen led another delegation to Moscow in July 1957. Under an agreement termed the *New Defence Technical Accord 1957–87*, signed 20 August 1957 but not ratified until 15 October, the Soviet Union agreed to supply missile models, technical documents, designs and specialists. In January 1958 the first of several R-2s reached the Fifth Academy under cover of darkness. A hundred Soviet specialists arrived, bearing over 10,000 blueprints and technical documents. The first group of 50 Chinese graduates went to Moscow for study. While there, some zealous Chinese students at the Moscow Aviation Institute engaged in some freelance espionage on behalf of their country, copying and stealing restricted information. However, despite a much more intense level of cooperation, the Chinese felt all along that the Russians were being less than wholehearted. Requested information was often not supplied or was referred to higher authority, where the request disappeared into a bureaucratic black hole.

PROJECT 1059: THE CHALLENGE OF COPYING A RUSSIAN ROCKET

In January 1958, the Fifth Academy adopted the *Essentials of a ten-year plan for jet and rocket technology, 1958–67*. With the Soviet Union making rapid strides in space exploration (a probe was even sent to the Moon early in 1959), Chinese scientists were divided as to the wisdom of working on rockets so obviously out of date as the Russian R-2. The Fifth Academy took the view that even to copy the R-2 would be a difficult enough challenge in itself, but they decided to go ahead, and named the enterprise 'project 1059', beginning a tradition of numerical allocations to keynote projects.

Table 1.1. The coded projects of the Chinese space programme

331	–	communication satellite, 1977.
581	–	plan for an Earth satellite project, 1958, renewed in 1965 as project 651.
701	–	Ji Shu Shiyan Weixing technology satellite, 1970.
761	–	new sounding rocket programme, 1977.
911	–	recoverable satellite programme, 1967.
921	–	manned spaceflight programme, 1996.
1059	–	to copy and fire a Russian R-2, 1958.

China lacked the ability to bring together the key material elements of any rocket – even the most basic ones. Throughout early 1959 there was continuous traffic between Russia and China as the Academy tried to get the R-2 into production. Two Chinese delegations were in Moscow that August to ask about additional equipment. During the war, production of the A-4 had consumed a considerable part of Germany's war effort, using scarce supplies of steel, aluminium and rubber at a crucial stage of the war and requiring huge numbers of hours of assembly. The Chinese were now learning the hard way that production of a rocket was a demanding and sophisticated task. Under the Chinese–Soviet accord, an Academy of Sciences delegation had visited Moscow in October 1958. Formally titled a High-altitude Atmospheric Physics Delegation, under the leadership of Zhao Jiuzhang, Wei Jiqing and Yang Jiachi, the tour quickly became aware of the level of complexity and organisation required to run a rocket programme.

THE IDEA OF A CHINESE EARTH SATELLITE

It is not known how prepared the Chinese were for the launch of Sputnik 1 (the Soviet media had been publicly announcing its proposals to launch an Earth satellite since 1954, but no-one in the West had paid much attention). When Sputnik 1 was launched on 4 October 1957, the Chinese Academy of Sciences set up an optical observation administration office, and observation stations were built in Beijing, Nanjing, Guangzhou, Wuhan, Changchun, Yunnan and Shaanxi. Their observations were coordinated in the Zijin Shan Purple Mountain observatory of Nanjing, which dated from 104 BC and which now followed Sputnik's path around the Earth and predicted its next passes.

Tsien Hsue-shen had already proposed an Earth satellite during the period of the development of the Chinese R-2 rocket. At the second meeting of the VIII Communist Party Congress, Mao Zedong approved the idea. On 17 May 1958, apparently encouraged by the performance of the Soviet Union in orbiting a large, 1.5-tonne satellite (Sputnik 3) two days earlier, he told his colleagues that China too must launch Earth satellites. His only rider was that a Chinese satellite should be large and not the size of the small American satellites then being put into orbit (Vanguard 1 was a tiny 1.5 kg). In August 1958, the state Scientific Planning Commission endorsed Mao's proposal, arguing that a satellite

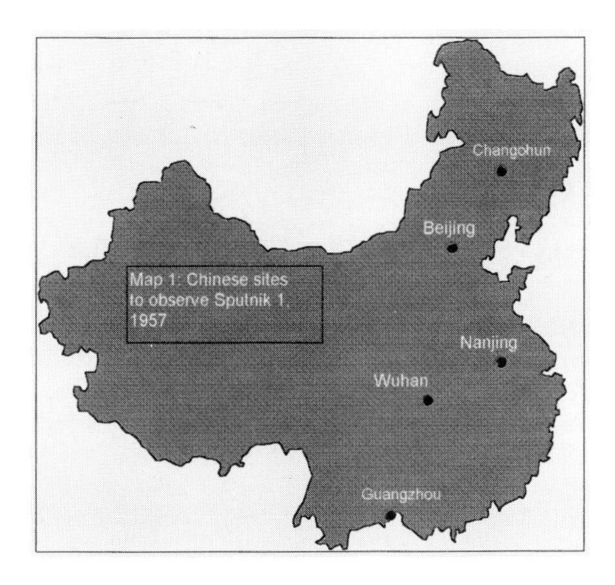

Map of Chinese sites to observe Sputnik 1.

would mobilise military rocket development, impress foreign powers and spur the development of other areas of science and technology. Nie Rongzhen, as deputy premier, called a special meeting of the Fifth Academy and asked Tsien to get together the best people to draw up a plan for the development of the satellite.

> Mao Zedong, 17 May 1958
> '*We too must build artificial satellites*'

PROJECT 581

The Earth satellite was given a project code number (project 581) and made a priority national task. A '581 group' was brought together, soon to be divided into teams for engineering (institute 1001), control systems and research tasks. In October, the Academy of Sciences organised a scientific exhibition, which included models of rockets and satellites. Visitors included Mao Zedong and Zhou Enlai. In November 1958, the task of building the satellite was given to the Shanghai Institute of Machine and Electricity Design (SIMED), which was under the dual management of Shanghai municipality and the Chinese Academy of Sciences. The 1001 team was transferred to work there under the guidance of Prof. Yang Nansheng and Prof. Wang Xiji, and young engineers were drafted in from Shanghai's industries and engineering centres, which were the most advanced in China.

There were long delays getting the project under way which have never been fully explained. Mao's support came in the middle of the great leap forward, when a period of rapid modernisation in industry, agriculture and the economy had been announced, including the bizarre national campaigns to eliminate rats, flies, mosquitoes and sparrows. These campaigns – in which all citizens were expected to participate – certainly disrupted the space effort, but worse was to come, in the autumn, with the drive for steel production, in which every home and factory was required to melt steel in order to increase national output. By the following year the economy was collapsing and millions were starving, and even the relatively pampered senior scientists around Tsien went hungry. At one stage, 70 per cent of those working on the project were suffering from oedema due to malnutrition. Marshal Nie Rongzhen intervened, and ordered that they be supplied with rations from the navy.

EARTH SATELLITE SHELVED; ROCKET PROGRAMME CONTINUES

The idea of an Earth satellite was shelved temporarily in January 1959, the decision being taken to concentrate on sounding and military rockets, and to drop the satellite project. The view was taken that China did not have the resources to do both[3]. The Chinese lacked vital skills in the areas of machining, tooling, welding, pressing and punching. Facilities for rocket production were primitive and had to use converted aeroplane hangars. The directive was issued by party Secretary Deng Xiaoping, and the three teams were broken up. Many years later, the decision was summarized as follows:

> In January 1959, Deng Xiaoping issued an instruction to the effect that, in the current situation, when the national resources were not available for launching satellites, they should readjust their space technology research duties. The

> party group of the Chinese Academy of Sciences called a meeting to study the question of the fundamental policy for their work to develop artificial satellites and came to the conclusion that it was necessary to correct the erroneous tendency to be over-anxious to have artificial satellites in circumstances where the basic conditions for achieving this were not currently present[4].

At one stage, there was considerable pressure to also abandon the rocket programme (now coded project 1059), since it was too advanced for China's state of development at the time. Matters came to a head at the National Defence Industries Conference held in Beidaihe in July 1961. Mao Zedong and Zhou Enlai insisted that the rocket project press on.

Substantial numbers of personnel were requisitioned for project 1059. Beijing council assigned 6,000 of its workers to building work, and at one stage, 1,400 institutes and enterprises and 14 factories were involved. By 1961, the workforce of project 1059 had swelled to 15,000 men working on sites totalling a million square metres. Top graduates were assigned to the project as their first work, and the army sent its top leadership cadres. Zhou Enlai insisted on regular progress reports. Some of the items necessary for the project, such as aluminium alloy plate, seamless stainless-steel tubes and some rubber items, were simply not available in China. Up to 40 per cent of the Chinese R-2 used substitute materials made in their stead. Some technical skills, especially in welding, were poorly developed and required intensive Soviet help. Those involved in the project worked in some very difficult conditions, in factories which were unheated in winter and unventilated in summer. Many slept in their workplaces by night and took their meals at their workbenches in order to save time. Although some had dormitory accommodation, others lived in tents. Some contracted dropsy as a result of the poor conditions.

A key personality in the development of Chinese rocketry in this period was Tu Shoue (b. 1917). Like Tsien, he had a background in the United States, having enrolled at MIT in 1941, and worked subsequently in the Curtis aircraft plant in Buffalo. Returning to China, he became professor at Qinghua University and subsequently Beijing. From 1957, he was drafted in as head of design in the Fifth Academy and led the development of the R-2 and military rockets in the late 1950s and early 1960s. One of his leading colleagues was Liang Shoupan (b. 1916), a 1939 graduate from MIT, professor of aviation engineering in China during the war, and brought into the rocket programme in 1956.

DEVELOPMENT OF A SOUNDING ROCKET

Independently of the main effort with the R-2, the Shanghai Institute of Machine and Electricity Design began in 1959 to develop a sounding rocket called the T-5. Sounding rockets are small rockets used to probe the atmosphere, the most popular use being for weather forecasting. They are flown on up-and-down missions of short duration and are not intended to reach orbit. Although the scale involved in designing and firing sounding rockets is much less than that of an orbital rocket, the basic principles of rocketry are the same. Granted the complete lack of experience of the Institute, it was an ambitious project for its time, but one in keeping with the lofty idealism of the great leap forward.

Designed to study the geophysical phenomena of the atmosphere, the T-5 was built and assembled in 1958. However, the young engineers were unable to obtain the materials necessary to construct an engine test-bed, or, more seriously, obtain liquid oxygen, and they had to abandon the project a year later. Instead, in October 1959 they decided to develop a smaller, simpler sounding rocket – one more tailored to the country's tight economic circumstances. This was the T-7, but it was to be preceded by a development version, the T-7M. This was a small test rocket with a liquid-fuel stage and solid-fuel stage. Possibly because of the inability of Chinese industry to produce and store liquid oxygen, the T-7M used storable fuels (in this case, nitric acid as oxidiser, and anilene and furfuralcohol as fuel). It weighed 190 kg, was 5.3 m tall, had a thrust of 226 kg and was designed to reach up to 10 km. Unlike the T-5, it was unguided, with four triangular tail fins. Chief designer was Yang Nansheng and chief engineer Wang Xiji. Their progress was inspected by President Liu Shaoqi and General Secretary Deng Xiaoping that December.

The T-7M's first launch took place at Laogang, outside Shanghai, on the East China Sea, on 19 February 1960. Launching facilities were crude. The team calculated trajectories with a hand computer, worked in an old airport shed, and turned the tracking antenna by hand. A bicycle pump was used to pressurise the fuel tank! On its first mission, the T-7M headed into the clouds, reaching a height of 8 km. That May, the rocket was put on display at the Shanghai New Technology Exhibition and was inspected personally by Mao Zedong.

After the success of the T-7M, work began at once on the operational version, the T-7. The first hot engine run of its motor took place on 18 April 1960 on a site set aside at Shanghai airfield, and was watched by Nie Rongzhen and Tsien Hsue-shen. In the meantime, it was decided to set up a new launch site for the programme. The Academy of Sciences, with Shanghai municipality, set up a new site at Shijiedu, inland in Anhui province in the hills of Guangde.

Shijiedu was ready for the operational version, the T-7, to make its maiden flight in September 1960. The T-7 weighed 1,138 kg on the pad and reached an altitude of 60 km. It carried a payload of 25 kg of meteorological sounding equipment, designed to measure the atmosphere's pressure, density, temperature, wind speed and direction. The guidance engineer was Zhu Yilin.

LAUNCH SITE IN THE GOBI DESERT

Even as the Shanghai team made progress with sounding rockets, events were moving to a climax with the development of China's first 'real' rocket. By this stage, good progress had been made in building China's first launch site, with Soviet advice. The Russians had already constructed three rocket launch sites: Kapustin Yar, on the Volga, where they had tested the German A-4s; Baikonour, or Tyuratam, in Kazakhstan, where they had launched the first Sputnik in 1957; and Plesetsk, northern Russia, which was their first intercontinental missile base. For China, one of the top priorities was to site the base away from the coastal regions which were most vulnerable in the event of conflict.

The Chinese launch site was settled at Jiuquan, 1,000 m high in the Gobi mountain desert, in Gansu province, 1,600 km west of Beijing and on the ancient silk road. Also known as Shuang Chengzi base, like Baikonour it was selected on account of its low population density (in the event of rockets failing) and remoteness (for secrecy). A railway line was connected from the Fifth Academy in Beijing to the launch site in Jiuquan via the Lanzhou–Urumqi railway (which must have been a considerable journey for the rocketeers, as travel there takes five days). The 300-km branch line starts at Qingshui town, 40 km east of Jiuquan city, and heads northward along the Ruoshui river, ending at Saihan Toroi, inner Mongolia.

In April 1958, the 20th Corps of the Chinese army was ordered to the area to construct the launch site. Sun Jixian was appointed first commander of Jiuquan, with Li Zaishan its first commissar (his job was to ensure political discipline). Conditions must have been harsh then: bitterly cold in winter and baking hot in summer, exacerbated by strong winds. Water had to be brought in by truck while the first wells were drilled. Zhou Enlai came to investigate progress. Before long, construction was noticed and photographed by prying American reconnaissance satellites[5].

THE GREAT SPLIT

The 30-year unshakeable Chinese–Soviet accord ended in tears in August 1960. No single reason explained the breakdown in relationship. The refusal of the Soviet Union to supply specifically requested nuclear technology appears to have been the primary cause, Khrushchev apparently becoming more and more convinced that China had every intention of using it in a nuclear war at the first available opportunity. On 12 August 1,400 Soviet technical advisers returned home abruptly, taking their blueprints with them and shredding anything they could not carry, and more than 200 joint projects were cancelled. The departure of the Soviets from the Fifth Academy was apparently good-natured and an occasion of genuine regret for both parties. Photographs taken at the time show the scientists bidding each other fond farewells.

The split between the two communist allies came at a crucial moment. In September 1960, the Chinese made sufficient progress to launch a Soviet-supplied R-2 rocket, though using Chinese fuel. The real test was not far off. The construction and launch teams laboured in their factories and in the Gobi desert throughout October.

Director of the impending crucial test was Marshal of China, Nie Rongzhen. Two months later, on 5 November 1960, China launched its first real rocket actually made in China, and Tsien himself was there to watch the rocket brought down to its pad, lifted into place, fuelled and fired. Nie Rongzhen presided. As the R-2 disappeared over the horizon, the celebrations began. Two more R-2s were successfully fired the following month. Although hardly anyone outside its scientific and political leadership knew it, China had joined the space race. The Chinese R-2 was named the Dong Feng 1, or 'East Wind' 1, and the success was publicly reported in December.

Leaders of the launch of the Chinese R-2, 5 November 1960

Responsible for the strategic rocket programme: Marshall Nie Rongzhen
Responsible for rocket development: Tsien Hsue-shen
Organiser of first test flight: Col Gen Zhang Aiping
Responsible for control: Wang Zhen

For four nations, the A-4 rocket had marked the coming-of-age of their rocket programmes. The first A-4 had been fired into the atmosphere over the Baltic by the Germans on 3 October 1942. The Americans had copied the A-4 at White Sands in New Mexico in 1946, the Russians at Kapustin Yar the following year (the R-1). Now the Chinese too had emulated the German, Soviet and American achievement. Would China now be the world's next great space power?

ASSESSMENT AND CONCLUSIONS

China joined the space race on 8 October 1956, even before the first Earth satellite had been launched. The Chinese saw rocketry and space science as an important part of national defence, in leading industrial and technical development, and in developing an economy in the full flush of socialist construction. China's most talented scientists were expelled from the United States at the very moment that the Chinese political leadership at home was anxious to modernise the country. China unexpectedly and suddenly inherited many of the best scientific experts in the world.

In the heady, early days of socialist construction, it seemed possible that China could soon build an intercontinental rocket *and* an Earth satellite, possibly joining the Soviet Union and United States in space as a close third. However, despite their great theoretical knowledge, the Chinese soon realised that their reach exceeded their grasp. The cruel realities of an underdeveloped economy struck home, and they soon realised that the construction of a basic rocket, to the standards that the Germans had achieved in the early 1940s, would take much time, effort and patience. In pragmatically turning to the Soviet Union, the Chinese were able to receive some grudging, then more forthcoming, help and assistance. It was only when Chinese scientists toured the Soviet Union that they realised the considerable industrial and technical base that was necessary to underpin the building of modern rockets and an Earth satellite.

Despite the enthusiastic start of projects to reverse-engineer a Russian rocket (project 1059) and build an Earth satellite (project 581), Chinese ambitions unravelled. The Earth satellite project fell victim to the great leap forward and the dawning and unpalatable realisation that this project was one bridge too far. The Sino–Soviet split almost put paid to the project to fire the R-2. Thankfully for the Chinese, it was so advanced that they were able to complete the R-2 within months of the Soviet departure. But with the Russians gone and China embargoed by the western world, they must have then realised that they were on their own.

2

Dong Fang Hong – the East is Red (1960–1971)

China's initial success with the R-2, or Dong Feng 1, belied the difficulties which lay ahead. The development of the Dong Feng series of missiles in the 1960s took much more time and effort than was anticipated. It was some time before the idea of the Earth satellite was to be restored to the agenda: even when it was, progress was hesitant. Concentration on military needs came first.

MILITARY IMPERATIVES: DONG FENG 2

As the Earth satellite project faded into the background, military imperatives came to the fore. While western perceptions of China have been traditionally of an aggressive, insurgent communist Asian troublemaker, China for its part felt isolated, friendless, surrounded by powerful enemies (principally the United States) and betrayed by its former friend, the Soviet Union. For these reasons, building up missile defences assumed the upper hand, rather than satellite projects.

The next rocket, Dong Feng 2, would be China's first indigenously designed and manufactured rocket, with a range of 1,500 km, capable of hitting the old wartime adversary, Japan. It would match the Soviet ballistic missile of the mid-1950s, the R-5. Dong Feng 3 would reach the Philippines (10,000 km), Dong Feng 4 the mid-Pacific, and Dong Feng 5 would be an intercontinental ballistic missile to match the American Atlas and the Soviet R-7 used to launch Sputnik. Further in the future, China planned to match the ability of the United States, the Soviet Union, France and Britain to launch missiles from submarines.

Denied Soviet assistance, the Chinese found the development of an indigenously designed and built missile to be cruelly difficult. The first attempt to launch Dong Feng 2 ended disastrously when it shook violently and crashed 69 s into its mission, on 21 March 1962, at Jiuquan. The post-mortem found that the engine had not been structurally installed correctly, that the gyroscope was in the wrong position, that the guidance system was defective, and that the rocket frame was unable to withstand elastic vibration. The project leaders were mortified by the failure, but Nie Rongzhen rallied them and told them to apply the lessons so that this problem never arose again.

Tsien set to work to redesign the guidance system. The redesign took over two years, involved 17 ground tests and did not achieve success until 29 June 1964. Further successful launches were carried out on 9 and 11 July. Although officially and politically this was a period of 'adjustment, consolidation, replenishment and improvement' after the great leap forward, the rocket project continued to be a high priority. Resources streamed into rocketry and related technologies (e.g. computers). During 1962–64, work began on rocket engine test stands, a vibration testing tower, vacuum chamber, wind tunnels and solid-fuel motors. More constructively for the space industry, the role of the political commissars in the Fifth Academy was downgraded and the amount of working time devoted to party activities reduced to less than one-sixth. Subsequent official histories refer to this as the period in which the erroneous leftist methods of the great leap forward were rectified. In November 1962, on the proposal of President Liu Shaoqi, the rocket effort was brought directly under the Central Committee of the Communist Party of China, reporting directly to one of its committees.

CHINA A NUCLEAR POWER

In 1964, Chinese nuclear science achieved its long-awaited breakthrough when in October its first nuclear weapon was exploded, during the same week that, incidentally, the foe Khrushchev was deposed in the Kremlin. Tsien and his team modified Dong Feng 2 so as to increase its range and give it an internal computer guidance system. Retitled the Dong Feng 2A, it was launched on 27 October 1966 on a 640-km westward path, carrying a live 1,290-kg nuclear weapon which, on impact, duly exploded in the Xinjian desert, releasing a 12-kton nuclear explosion. Chang calls this test the most reckless nuclear experiment in history. She may well be right, though only a few years earlier the Soviets had exploded warheads from rockets, and the United States had let off nuclear bombs high in the atmosphere. The Chinese nuclear tests in the 1960s had an approach to safety not that dissimilar from the American tests in Nevada a decade earlier. Photographs show the scientists and workers gathering to watch the atomic blasts in ringside seats at what seems to be only a few miles from the blast, before eagerly rushing forward to admire its effects. Radiation dangers seemed but a minor consideration.

The Chinese nuclear test led the American media to rediscover Tsien. They ran a series of articles deploring the way in which he had been expelled from the United States and lamenting that he now headed up the rocket and nuclear programmes of one of America's political adversaries. Ten years earlier, the American press had passed over his deportation in silence. Tsien was not the only American-trained nuclear expert, for there were others, including Zhu Guangya (b. 1924), who studied nuclear physics in Michigan from 1946 to 1950 and subsequently returned to head up the nuclear research programme; hydrogen-bomb expert Deng Jiaxian (1924–86), who studied at Purdue University from 1948–1950, and metal physicist Chen Nengkuan (b. 1923), a veteran of Yale, Johns Hopkins and the Westinghouse Institute. Several other leading atomic physicists, trained in other western countries, included Wang Ganchang (b. 1907) who trained in Germany in the 1930s and Russia in the 1950s; and Cheng Kaijia (b. 1918), who studied at the University of Edinburgh, Scotland, from 1946–50.

DONG FENG 3 AND 4

The next project, Dong Feng 3, had originally been for a 10,000 km range missile, but had to be suspended due to a series of technical setbacks. The design was simply too ambitious for Chinese manufacturing capabilities at the time, but the project eventually got under way in early 1965. There were two important innovations in Dong Feng 3. First, the rocket clustered four engines together at take-off, not just the single engine of the Dong Feng 1 or Dong Feng 2. Such clustering required the four engines to achieve virtually the same level of thrust simultaneously, otherwise the rocket would go off course. Second, the Chinese abandoned the fuels used on the Dong Feng 1 and Dong Feng 2 for storable propellants. Storable fuels had the handling advantage that they could be kept in the rocket for a long time before they were launched (important in the period of long and troublesome countdowns) and the related military advantage that such rockets could be fired speedily. The principal disadvantage is that storable propellants, which use nitric acid as oxidiser and either nitrogen tetroxide or UDMH (unsymmetrical dimethyl hydrazine) as fuel, are highly toxic, corrosive and environmentally damaging. The dangers of corrosion required the engineers to use new metals in building the tanks, seals and motors. Furthermore, they used new forms of high-strength aluminium alloys 50 per cent stronger than those previously used, enabling a reduction of the rocket weight and an increase in the payload weight.

Dong Feng 3 was the first Chinese rocket subjected to intensive ground testing; the four engines were first tested together on a stand in July 1965. As rocket engines became more powerful, the engineers encountered new problems, such as unstable combustion caused by vibration. On 30 July 1966 Zhou Enlai watched a countdown rehearsal during a visit to Jiuquan, where Dong Feng 3 eventually made its first flight on 26 December 1966. The next batch of tests went well, and the missile was considered operational in 1969. (It was named CSS-2 in the West, and a descendant was eventually developed and sold to Saudi Arabia).

Dong Feng 4 was an improved version of Dong Feng 3. It came to have a dual identity. With a small upper stage, the civilian version of Dong Feng 4 became best known as the historic Long March 1 rocket which launched China's first satellite. It made its first flight on 30 January 1970 and was subsequently formed into a missile strike force in Harbin, capital of Manchuria, in the north-east.

DONG FENG 5: CHINA'S LONG-RANGE ROCKET

Dong Feng 5 was important for the development of Chinese Earth satellites and for the Chinese military. Russia had built an intercontinental ballistic missile first (the R-7, August 1957) and then immediately turned it into a satellite launcher (Sputnik 1, October 1957). For their first successful satellite, the Americans uprated a medium-range rocket (the Jupiter), attached an upper stage and put Explorer 1 into orbit (January 1958). America's intercontinental ballistic missile (Atlas) came later in 1958. In essence, the Chinese followed the American approach, using a medium-range missile (Dong Feng 4) as the basis for their first satellite launcher (Long March 1), then building an intercontinental ballistic missile later (Dong Feng 5). Chinese space histories give considerable attention to Dong Feng 5, for with it, the Chinese achieved military rocket comparability with the United States and the Soviet Union. An ICBM could now strike the United States, China's most likely superpower enemy.

Of equal significance, the work done perfecting Dong Feng 5 was passed on to two rocket teams – the Beijing team designing Long March 1 to launch China's first Earth satellite, and the Shanghai team designing a successor set of military satellites on their own launcher, the Feng Bao ('storm'). Dong Feng 5 required the Chinese to make considerable progress in guidance systems, propulsion, rocket engines, steering devices and strong but light materials. Its chief designer was Tu Shoue.

Work on the Dong Feng 5 came to a virtual standstill in 1967–69 because of the Cultural Revolution and the switching of people and resources to the work on the Long March 1 rocket, and Dong Feng 5 scientists and technicians were assigned for re-education, land reclamation or manual labour. The first flight of Dong Feng 5 eventually took place from Jiuquan on 10 September 1971, and a base was later established at Harbin, in Manchuria, from where early test launches were made, to impact in Tibet and the Taklimakan desert in Xinjiang.

At this stage, Dong Feng 5 had been tested only on short-range missions, within China. Its long-range ability was untested. A comprehensive round of tests was approved in 1977, with further flights from Jiuquan to within China in 1978–79. These were successful, and it was decided to then fly the rocket to its full range, height and speed. This would require a trajectory far outside China's borders. Dong Feng 5 would be launched to an altitude of 1,000 km, fly at 7 km/s, impact in the southern Pacific half an hour later and reach a range more than 9,000 km distant. Its impact would be observed by China's ocean-going communications ship (comship), the *Yuan Wang*. The main problem in preparing the launching was the unreliability of the electronic components, which repeatedly gave trouble. Deputy premier Zhang Aiping therefore set up a trouble-shooting group (called the Electronic Component Reliability Work Leadership Group) to remedy it, with evident success.

On 26 April 1980, the new comship *Yuan Wang 1,* following inspection by Zhang Aiping and representatives of the state council and the Central Committee, was seen off from its port in Shanghai to monitor the impacts in the South Pacific. Two weeks later the fleet arrived on station, close to the equator, about 1,000 km from the Solomon Islands. In the early dawn of the day of the test, 18 May 1980, the wind had dropped, the waves lapped gently around the ships and numerous clouds drifted past.

Back in China, in the still night-time morning of 18 May 1980, Tu Shoue and his colleagues gathered around Dong Feng 5, bathed in floodlights. The atmosphere was tense as 10 years of work reached the moment of truth. Dong Feng 5 roared off into the clouds, bending over in its path across China towards the Pacific. Ten tracking stations reported its arc across the morning sky. At sea, sailors observed a huge ball of fire, scattering into bright spots, as the rocket crashed Earthward into the upper atmosphere. However, they could note one bright spot still falling and burning its way through re-entry. With a sonic boom and a whistling noise, a parachute opened and the recovery cone made a tremendous 100 m high splash. A helicopter took off at once from one of the recovery ships, dropping diver Liu Zhiyou into the sea to attach a flotation system. The recovery cone was hoisted on board in 15 minutes.

Another Dong Feng 5 payload was observed impacting on 21 May by prying ships of the Australian navy. On 2 June, *Yuan Wang* and its attendant ships returned in triumph to port in Shanghai. There was joy throughout the rocket community in China, Zhang Aiping even being moved to write a poem to mark the event. A week later, all those involved in the Dong Feng 5 tests were received by Deng Xiaoping and Hu Yaobang in the Great Hall of the People in Beijing.

SUBMARINE-LAUNCHED MISSILE

Progress on the submarine-launched missile was slow. Because submarine-based missiles might spend long periods at a time at sea before their launch, they had to be solid-fuel rockets. The Chinese decided on a powerful two-stage submarine-launched ballistic missile. The solid-rocket motor nozzles were designed to be swivelled to ensure accurate guidance so that the rocket hit the correct target. This proved to be a knotty problem which took years to resolve. The programme was impeded by the Cultural Revolution; and worse, on 16 March 1974 technician Wang Lin, deputy head of the rocket workshop, was unable to rest that night as he contemplated new ways of mixing more effective propellant. He returned to the workshop. Later that night, the mixer exploded, and Wang Lin was buried underneath the protective iron door of his laboratory, where he died.

The first and second stages underwent 10 successful ground runs in the late 1970s, and made three land-based launches in 1981–82. A further disaster – this time a natural one – intervened in August 1981, in the final stages of assembly of the rocket for the first underwater test. The factory which made the gyro platform found itself, ironically, under water itself when flash floods tore down from the mountains, destroying roads and parts of the factory. The staff disassembled the gyroplatforms, carried them in backpacks across mountains and delivered them to the nearest flood-free railway line.

Two years later, on 12 October 1982, a Chinese navy submarine left harbour and dived close offshore. After a few minutes, it launched its first submarine-fired ballistic missile, arcing out over the sea, and coming down in the distant ocean. China thereby demonstrated that the country had a land- and sea-based operational intercontinental ballistic missile system at least on a par with France and Britain. Designer of the submarine-launched ballistic missile was Huang Weilu (b. 1916), a 1940s graduate of the University of London who subsequently went on to be one of China's main pioneers of radio control and guidance systems and designer of solid rockets.

CHINA'S ROCKET STRIKE FORCE

Ultimately, Dong Feng 4 and 5 were developed as China's nuclear strike force. Dong Feng 5 was known as the CSS-4 by NATO and carried a hydrogen nuclear bomb weighing 3 tonnes. By the late 1990s, China had up to 20 Dong Feng 4s (able to reach Moscow) and between seven and 15 Dong Feng 5s, able to reach the United States, a small but potent arsenal[6]. They served thus until the late 1990s, when they began to be phased out by Dong Feng 31 (range 8,000 km, also submarine-launched) and Dong Feng 41 (12,000 km) respectively. These were solid-rocket missiles, broadly equivalent to the Russian SS-25 or the American Minuteman, and were set for location in two secret silo-based centres. They could strike not only the American west coast, but inland as far as the Rockies[7]. American analysts were divided as to whether this signalled purposeful modernisation of an obsolete military apparatus or a more aggressive foreign policy[8].

Table 2.1. China's missiles, the Dong Feng series

Missile	Date launched	Notes
Dong Feng 1	5 November 1960	Reverse-engineered Russian R-2, in turn based on German A-4.
Dong Feng 2	29 June 1964	Designed to reach Japan, range 1,500 km.
Dong Feng 3	26 December 1966	Designed to reach the Philippines, range 10,000 km.
Dong Feng 4	30 January 1970	Medium range, designed to reach the mid-Pacific. Lower stages were the basis for Long March 1.
Dong Feng 5	10 September 1971, operationally on 18 May 1980	First Chinese intercontinental ballistic missile. Basis for Long March 2 and Feng Bao.
Submarine-launched ballistic missile	12 October 1982	
Dong Feng 31 Dong Feng 41	Introduced in 1990s	Range: 8,000 km, submarine- and land-launched Range: 12,000 km, land-launched.

SOUNDING ROCKETS MAKE PROGRESS

Even as China was making progress with the development of its military missiles, the sounding rocket teams were paving the way for China's eventual entry to the space race. In 1962, the Academy of Sciences put forward the specification for a more advanced weather sounding rocket, better than the T-7. Designer Wang Xiji responded by improving the tail fins, reducing the weight of the engine, increasing the thrust, adjusting the upper stage to work at altitude and stretching the length of the tank. The new rocket was the T-7A, which weighed 1,145 kg and stood 10.32 m tall. It was first launched in December 1963, carried 40 kg of equipment and reached an altitude of 115 km.

The T-7 and its improved version, the T-7A, were used many times in the 1960s for meteorological, geophysical, technology development and biological missions, though, as with similar Soviet missions, a full flight log does not appear to be available. The meteorological version first flew in December 1963. Scientific versions were used to survey the ionosphere, electron densities, cosmic rays and the Earth's magnetic field.

BIOLOGICAL SOUNDING ROCKETS

Under the management of the Biophysics Research Institute of the Chinese Academy of Sciences and the Shanghai Machinery and Electrical Equipment Design Academy, the

T-7A was used for a series of biomedical missions in the course of 1964–66. The series was called the T-7A-S. The nose of the T-7A was redesigned to take a sealed biological cabin with a standard cargo of four white rats, four white mice and 12 biological test tubes which in turn would hold fruit flies and other test items. The aim was to observe, with a camera, the behaviour of the animals in flight. After the flight, some of the animals would be dissected to see whether there had been any effect on their biology; the others would be bred to watch for genetic change. The T-7A-S had a weight of 1,165 kg and a height of 10.81 m.

On its first mission, the biological sounding rocket took off on 19 July 1964, reaching an altitude of 70 km. Film was taken of the reaction of the cargo to weightlessness. On 1 and 5 June 1965, two further missions were carried out. The animals were recovered intact on all three occasions.

DOGS FLY INTO THE ATMOSPHERE: THE MISSIONS OF XIAO BAO AND SHAN SHAN

The Academy of Sciences decided at this stage to proceed with flights of dogs. Canines had been the main precursors of humans in the Soviet space programme. The T-7 was adapted a second time, with a larger sealed cabin, able to take a dog, four white rats and 12 biological test tubes. The resulting sounding rocket, the T-7A-S2, weighed 1,346 kg, of which the biological container constituted 170 kg. It was 10 m tall, with a diameter of 0.45 m, and was designed to reach between 60 km and 115 km. The purpose was to test the reaction of the dogs under weightlessness, ascent and descent, and their reactions to noise and vibration. The level of radiation dosage was also to be tested. During the flight, the dog's heart-beat, temperature, respiration and breathing rates would be registered by a tape recorder. The carrying of a dog required a much more advanced life-support system, but as a precaution against a delayed recovery, arrangements were made for a pressure valve to be released during the descent to let in fresh air.

On 14 July 1966, the T-7A-S2 made its first flight carrying China's first space dog, Xiao Bao ('little leopard'). All went well. An Air Force helicopter spotted Xiao Bao's cabin drifting down, impacting at less than 10 m/s. The helicopter touched down alongside, and the dog's handler rushed forward to see Xiao Bao alive inside. When the handler took out the dog it was wagging its tail and was apparently glad to be back. It returned to kennels, but the rats were dissected on the spot. The second dog flight took place on 28 July 1966, this time carrying a bitch, Shan Shan ('coral'). It went equally well.

The T-7 was then adapted to carry satellite test equipment – satellite control systems, for example – that would later be used in orbit on China's first Earth satellites. The designer was Lin Huabao. Jiuquan was used as a launching site, with a spiral cage launch-rail system devised by Jiangnan shipyard in Shanghai. Two launches were made: one in 1965, reaching 83 km, the other in 1969, reaching 81 km. Instruments on board took pictures of the ground, photographs of the stars, obtained data on electrons in the ionosphere and tested out infrared scanners. The descent cabin used a mixture of airbrakes, braking parachute and main parachute to lower it to the ground at a speed of less than 6 m/s.

The first mission nearly led to disaster. The cabin came down in the desert in Badain Jaran. The recovery team of 20 people, led by Lin Huabao, set off in trucks but there were sand dunes everywhere which the vehicles could not negotiate. Searching on foot, they did not find the cabin until after nearly a day's searching in extremely sandy desert. Only the tip of the cone was showing, the rest being buried in dunes. It took them three hours to remove the scientific and test equipment – whatever they could carry by hand. In their attempt to return to their trucks that night, they lost their way, but although they fired flares, nobody saw them. They had a terrible night, having run out of water and with their feet blistered. Lin Huabao rallied the team and they managed to crawl out of the desert the following day.

RESUMPTION OF THE EARTH SATELLITE PROJECT

Project 581, the first Chinese Earth satellite, had been postponed indefinitely in 1959. The success of the R-2, the subsequent development of the Dong Feng military missiles and the successful flights of sounding rockets gave Tsien Hsue-shen the confidence to press for a resumption of work on plans for an Earth satellite. In the early 1960s, countries such as France and Britain began to develop launchers to put small Earth satellites into orbit, and it was clear that space would not remain the preserve of the two superpowers indefinitely.

In 1962 Tsien recruited a satellite development team of four men from the Shanghai Institute of Machine and Electrical Design (SIMED). They spent a year with Tsien in Beijing, working through possible designs and consulting the Russian- and English-language literature. The four were Zhu Yilin, Kong, Li and Chu. They seem to have had access to good information on a range of American space programmes such as Discoverer, Mercury and Ranger. The team met Tsien weekly to review progress, and by the end of the year they had drawn up a plan to cover the years 1964–73 for the development of Chinese Earth satellites. They also lectured students in the Chinese University of Science and Technology. While they were doing this, Tsien concluded his book *An introduction to interplanetary flight*, published in Beijing in 1963. This text brought together the sum of Tsien's knowledge of the subject to date, including the experience of developing the R-2, and also served as a basis for the lectures to the students.

Inspired by the flight of Yuri Gagarin around the Earth in April 1961, Tsien Hsue-shen convened a series of 12 symposia on space flight. The first was held in June 1961. The meetings continued for three years, during the course of which progress in spaceflight worldwide was reported and noted. The four designers acted as its secretariat. In 1963, the Academy of Sciences formed its Commission on Interplanetary Flight.

In January 1963, the Shanghai Institute of Machine and Electricity Design was made organisationally – on the order of the State Council – part of the Fifth Academy. When the four designers returned to Shanghai, they formed a satellite research division, and 50 technicians and experts were drafted into the process. They divided into five groups: overall design, structure and thermal control, attitude control, electrical power and tracking, telemetry and command. They were permitted to cooperate with other institutes in Shanghai, but not further afield.

A further spur to the development of the satellite came when the Academy of Sciences in Beijing established a committee to investigate the desirability of a man-made satellite. Its feasibility study began in May 1964 and concluded in July 1965. However, further authorisation for the project to continue was held up by the State Council. This caused much frustration in Shanghai, Tsien writing them letters of encouragement as he tried to grapple with the decision-makers above him.

PROJECT 651

Tsien sought authorisation for the resumption of work on an artificial Earth satellite from the Central Committee of the Communist Party of China in January 1965. He circulated a proposal, pointing out that an artificial Earth satellite had been on the agenda for seven years; that with the development of the Dong Feng 3 rocket progressing well, it should be possible to launch an Earth satellite; and that nothing was to be gained from waiting further.

Four months later, the party's Defence, Science and Technical Commission supported the project, recommending a satellite launch in the 1970–71 time period. The Commission proposed that the Academy of Sciences develop the satellite and the Seventh Ministry the rocket. The Academy responded to the invitation by presenting, in July 1965, *A proposal on the plan and programme of development work of our artificial satellites*. The academy proposed an evolutionary long-term development plan for China, starting with simple, experimental satellites and then progressing to more difficult tasks, applications programmes and the recovery of satellites. On 10 August 1965, Prime Minister Zhou Enlai and the Central Committee approved the plan, stipulating that the satellite should be visible from the ground and that its signals should be heard all over the world (similar considerations governed the approval of Sputnik 1). It was coded project 651.

REORGANISATION AND DISPERSAL

The decision to approve construction of the satellite laid down the lines of responsibilities between the different organisations and bureaux, and also commissioned the construction of naval tracking and communications ships (comships). Zhou Enlai took the opportunity to make further organisational changes to streamline and rationalise the space industry, concluding a process which had been under way for some years. The key elements of these moves are simplified here, though they were infinitely more tortuous and tangled than this abridged description indicates.

On 23 November 1964, the Central Committee of the Communist Party of China and the State Council issued a joint resolution restructuring the space industry. Their main purpose was to bring together the many different elements working on space activities in different parts of the country, in a dispersed and, they felt, uncoordinated way. All of them, including the Fifth Academy, were now transferred to what was called the Seventh Machine Industry Ministry. It was subdivided into four academies: long-distance rockets (1), ground to air (2), cruise (3) and solid (4), with sounding rockets and the bureau of tactical missiles located in Shanghai.

Whilst technically a change in name, in practice the change also meant a sharp change in status for the scientists and engineers. They lost military wages and were transferred to local wages, which were generally lower, and their food rations were reduced. In January 1965, Liu Shaoqi appointed Wang Bingzhang the first head of ministry, with Tsien as one of his six deputies. The Seventh Ministry quickly developed an eight year plan for the development of Chinese rocket technology, to cover the years 1965–72. Over 3,000 people in the ministry's design, production and user departments put forward their views. The new plan was adopted by the Central Committee in March 1965, and in May 1965 the new Seventh Ministry received a high-powered official visit from Liu Shaoqi, Zhou Enlai, Nie Rongzhen and Jian Qing (Mao Zedong's wife).

Responsibility for the satellite, which had been designed in Shanghai, was now handed over to a special bureau of the Chinese Academy of Sciences, called institute 651. This parallelled Soviet practices whereby the various design bureaux were given code numbers (Korolev's was OKB-1, Chelomei's was OKB-586, and so on). Not only that, but the Shanghai Institute was moved to Beijing, where it was renamed the Seventh Research Division of the Eighth Institute of Design in the Seventh Ministry of Engineering Industry, but was also known as the Beijing Institute of Spacecraft Systems Engineering (BISSE).

Whilst the transfer of the satellite project must have galled the Shanghai team, it nonetheless cooperated with the Academy of Sciences, and joint working groups were formed. The former Shanghai team and the Academy of Sciences institute 651 team formed one joint working group, the Chinese Academy of Space Technology (CAST), on 20 February 1968, and Tsien Hsue-shen was appointed the first President. All the existing institutes and bureaux working on satellite programmes were brought under the academy on the orders of Zhou Enlai. In a swop, several Beijing institutes were moved wholesale to Shanghai to form what subsequently came to be called the Shanghai base of the space industry (the Shanghai Academy of Space Technology (SAST)). The descendants of both bodies – CAST and SAST – survive to this day.

In November 1967 there was a further organisational change with the establishment of the Beijing Wan Yuang Industrial Corporation, later renamed the Chinese Academy of Launcher Technology (CALT), its current identification (although some Chinese literature still uses the older terminology). Thus, by the end of 1967 China had a unified academy of launcher technology and another academy of satellite technology, both of which persist to the present day.

A complicating theme of all of these moves was dispersal. Regarding the military threat to China at this time as very serious, the reorganisation specified in August 1965 that as many production units as possible should be shifted away from the coast, the south-east and the Sino–Soviet border, to as far inland as possible ('third-line regions').

64-DAY DESIGN CONFERENCE

The Chinese Academy of Sciences set up its own Artificial Satellite Design Academy in September 1965. An historic conference was held later that year, one bringing together all the groups concerned with the satellite project. The conference lasted 64 days – from 20 October to 2 December 1965 – and reviewed all the progress made so far. This was called the '651 conference' after the project code and the number of the institute which was building the satellite. The meeting was of the view that, although China had made a late start in the development of satellites, its first sputnik should be much more advanced than either the first satellites of the Soviet Union or the United States. It should be visible and audible to the Chinese people. The satellite would have a weight of at least 100 kg, fly out of its launch site in an easterly direction and enter orbit at inclination 42°. Sun Jiadong (b. 1929) was appointed chief designer. Unlike most of his American-trained colleagues, he was Soviet-trained, having spent the 1950s in the premier Zhukovsky Air Force Engineering College.

In May 1966 the satellite was designated Dong Fang Hong 1 ('the East is Red'). It was decided to increase the weight of the satellite to 170 kg, and it would fly to a much higher inclination (70°), which would mean that it could be seen and tracked over a much greater proportion of the Earth's surface. The decision specified that the visibility of the rocket be that of a fourth magnitude star (it was fitted with an extension to help) and the satellite that of a fifth magnitude star. The solar cells and scientific instruments were eliminated. Instead, a tape recorder would play stirring revolutionary tunes.

Dong Fang Hong 1 was a spherical polyhedron 1 m in diameter, made of fibreglass wound on aluminium alloy. The surface was plated with insulation material designed to protect the instruments from the extreme heat and cold of Earth orbit. A sealed instrument module contained five silver–zinc batteries able to provide power for 20 days. An electronic circuit was used to generate the tune 'the east is red'. On the outside, four 3-m short-wave antennas, four transmitting antennas and four transmitting beacons were fitted. Two sensors were installed: an infrared horizon sensor and a solar angle gauge. Five mockups of the satellites were produced, and these were subjected to tests of the design's ability to withstand heat and cold, to transmit, and to resist radiation and humidity. These tests were completed in October 1969.

At the same time, work began on the national tracking and control system. Originally, it had been intended to have a national control centre in Xian, Shaanxi, with seven observation posts scattered round the country, but Xian was not completed on time, and instead the task was undertaken by the launching centre in Jiuquan.

THE CULTURAL REVOLUTION ENGULFS CHINA

Again there were delays, and once again these were for political reasons. In March 1966 Mao Zedong launched the Cultural Revolution. By summer, the Seventh Ministry had come to a standstill. The Academy of Sciences, developing the satellite, was overrun by seething political factions. The ministry divided into two rival political factions, one favouring the Red Guards, the other the Liu and Deng Xiaoping clique. There were even armed clashes among the units responsible for building ground stations. Early in 1967 Tsien Hsue-shen was deposed and reduced to the status of an ordinary employee. Scientific approaches were ridiculed as revisionism, academics as reactionary and leading scientists as bourgeois intellectuals. Subsequent histories recorded that the revolutionaries spread the slogan that 'when the satellite goes up, the red flag goes down' and urged concentration on political rather than scientific tasks[9]. Titles in the workplace were abolished, chief designers being given no more authority over rocket design than the lowliest technician or support staff. Some scientists were locked up in cattle sheds, others banned from research and many sent to be 're-educated' in the countryside. Those who resisted were killed[10]. The rest bided their time, waiting for the bad times to pass. Some kept up their reading and scientific study clandestinely.

Zhou Enlai and Nie Rongzhen attempted to protect the satellite project from the ravages of the Cultural Revolution. On 17 March 1967, Zhou Enlai persuaded the Central

Committee and the State Council that the space and defence ministries and their associated projects should be brought under military control. The principal feature of military control was that such controlled areas were exempt from what the cultural revolutionaries called the 'big four' activities – free expression, free assemblies, big posters and the holding of great debates. Zhou Enlai issued instructions that the former leadership was to be reinstated, the missile programme was to remain a national priority and its leading scientists were to receive state protection. In April, the army moved in to enforce the decision. The new Chinese Academy of Space Technology (CAST) was first put under military law, so as to be an exempt body from the big four activities. It did not escape the revolution entirely, and its early work was impeded by the shortage of materials throughout the country. Units which had been transferred inland to third-line regions suffered acutely from revolutionary activity. But according to some subsequent Chinese histories, the space industry suffered less under the Cultural Revolution than many other parts of the Chinese economy[11]. The rest must have suffered grievously indeed.

ROCKET FOR THE FIRST EARTH SATELLITE

A new rocket was developed to support the project: Chang Zheng 1 (CZ-1), or Long March 1, named after the formative moment in Chinese communist history when Mao Zedong led his armies 8,000 km from the clutch of the nationalists into the distant but safe province of Shaanxi. Essentially, the Long March 1 was the civilian version of the Dong Feng 4 military missile, which in turn was an improved version of the Dong Feng 3. Design of the Long March 1 began in the Seventh Ministry, but during the reorganisation in November 1967 was switched to the Chinese Academy of Launcher Technology (CALT).

The specification was for a liquid propellant rocket able to send 200 kg into low Earth orbit. 30 leading engineers and technicians were assigned to the project. Ren Xinmin (b. 1915) was chief manager of the Long March 1 project, sorting out the principal problems of unstable engine combustion. From Ningguo, Anhui, he studied ordnance in college before going to the United States in 1945 and obtaining a doctorate in Michigan. Returning in 1949, he developed solid-fuel rockets for the People's Liberation Army before becoming director of the Liquid Propellant Engine Design Division of the Seventh Ministry of Machine Building. Outline design of the Long March 1 was completed within a month.

DESIGN CHALLENGES OF THE LONG MARCH 1

The first stage of the Long March 1 used nitric acid as oxidiser and UDMH as propellant; the second stage nitrogen tetroxide as oxidiser and UDMH as propellant. Designers of the first and second stages were Ren Xinmin, Ma Zuozin and Zhang Guitian. The Long March 1 design required the use of new alloys, such as titanium, glass-fibre reinforced plastic and new forms of high-strength steel. Novel techniques of advanced welding were used to complete the components to a high degree of perfection. For the first time, the first stage used a shared tank rather than two individual tanks. Whilst saving weight, this approach required high standards of plumbing, as any leakage would result in an immediate explosion. The engines for the Long March 1, whose design was also led by Ren Xinmin, involved new challenges. Four YF-2 engines were grouped together for the first stage. The engine for the Long March 1 proved to be extremely troublesome; it failed 17 running tests, mostly due to unstable combustion. The second-stage engine had to be adapted so that it could light at altitude (60 km) in low pressure and air density; accordingly, a special air evacuation chamber was built to simulate high altitude conditions and tested in November 1966.

Under the leadership of Huang Weilu, an inertial guidance system was developed for the Long March 1; a system much more advanced than the early Soviet rockets and up to the American standards of the mid-1960s. For electrical systems, the rocket marked the end of valves and the introduction of transistors, though this was far from trouble-free, owing to the poor production quality of Chinese transistors at the time. Two telemetry systems were fitted to the Long March 1 – one on the second stage, the other on the third – and these were designed to transmit back several hundred different parameters as the mission progressed. Finally, if the worst should happen, a destruct system was fitted. This could be activated either from the ground or by sensors on the rocket itself should it veer off course.

THE PROBLEM OF THE THIRD STAGE OF THE LONG MARCH 1

Dong Feng 4 had insufficient thrust to put a satellite into orbit, and for this reason the Chinese developed a small third stage with a solid rocket motor to enable the satellite to make the final leg of its journey into orbit. Designer of the third stage was Yang Nansheng. This was the first time that the Chinese had built a solid-fuel rocket on this scale. They had to start from scratch, as they had received no assistance from the Soviet Union in this area and other countries were embargoed by the United States from assisting China. The chemical industry had no experience of producing the kind of fuel which powers solid rocket motors, (the Americans use a form of perchlorate, which is poured into the rocket casing like a dark grey sludge).

The Chinese had recognised the need for solid rocket motor research when the space programme had begun in 1956. The Fifth Academy had included a solid propellant research group, but it comprised only three new graduates. By 1960, 70 people, led by Li Naiji, were working there. They went to some lengths to try to learn about solid rocket technology in other countries, picking up any engineering and industrial information they could. The Changchun Applied Chemistry Research Institute of the Chinese Academy of Sciences was commissioned to produce fuel – liquid polysulphide rubber; but the Harbin Military Engineering Academy produced the first batch of fuel – potassium perchlorate, and it was two years before a compound was produced of sufficient quality to burn. In 1959, the Fifth Academy teamed up with ordnance ministry factory #845 in Xian to produce the first significant quantities of pourable material. To take the process further, a national propellant conference was convened in January 1961, followed in March 1962 by the opening of the Solid-fuel Rocket Engine Research Institute in the Fifth Academy, of which Xiao Gan was the first head. Within a short period, the institute had developed a fuel pouring mixer and production line, and was conducting engine tests.

Then disaster struck. On 6 December 1962, 200 kg of solid fuel exploded in the mixer, killing two women technicians instantly. Two others died later of their injuries. Much more stringent safety systems were then introduced. It was the first of eight accidents in the Chinese space programme.

Table 2.2. Disasters and accidents in the Chinese space programme

Date	Event
6 Dec 1962	Explosion of solid rocket mixture in preparation, killing 4 technicians.
26 Jan 1968	GF-02 solid rocket motor explodes during tests.
16 Mar 1974	Solid rocket motor fuel mixer explodes, killing deputy head of workshop, Wang Lin.
Jan 1978	Explosion in third stage of Long March 3 rocket test engine, unspecified number of fatalities.
22 Mar 1992	Launch crews injured during on-the-pad abort.
2 Apr 1994	First geostationary meteorological satellite, Feng Yun 1, explodes during fuelling, killing one technician, injuring 31.
26 Jan 1995	6 villagers die, 23 injured, when Long March 2E carrying Apstar 2 explodes 70 s into mission.
14 Feb 1996	80 injured, between 2 and 56 fatalities, when Long March 3B crashes on maiden flight (also called the Saint Valentine's Day massacre).

In August 1964, a second national propellant conference was convened to review the (clearly slow) progress of the previous three years. Additional resources were devoted to attack the key problems which the institute was experiencing, such as the solid fuel cracking, the nozzle overheating and unstable combustion. There was progress when during the summer of 1965 six engines were fired successfully.

The specification for Long March 1 required a third, solid-fuelled upper stage 4 m long and weighing 1.8 tonnes. Its role was to fire at 600 km, when the second stage had completed its work, to put the Earth satellite into orbit. Preliminary design work was completed in April 1967, the institute opting for polysulphide rubber propellant, a high-strength steel casing made by Anshan Steel Works and a graphite nozzle throat liner. Their efforts at this crucial stage were complicated by the relocation of the academy to north-west China – a third-line region, where accommodation was very limited – and the Cultural Revolution. In order to have work completed on time, on more than one occasion in 1969 and 1970, the upper stage was declared by Zhou Enlai to be a State Priority Crash Construction Programme.

The first test run of the third stage, on 26 January 1968, was a failure, the engine exploding after 30 s due to a failure in the adhesive. However, by 1970, 19 tests had been carried out, five being high-altitude simulations. The one test that really mattered lay ahead.

The T-7 sounding rocket was used to test the solid rocket engine of the upper stage of the Long March 1 rocket, the carrier of the impending Chinese Earth satellite. The GF-01A solid rocket engine was fitted to the top of the sounding rocket in place of the biological or scientific payload. Two tests of the GF-01A were carried out, on 8 and 20 August 1968. The powerful third stage managed to boost the payload to an altitude of 311 km, which must have been the Chinese altitude record at the time.

LONG MARCH 1: THE FINAL LAPS

A priority area of work in the satellite project was the building of a rocket engine testing site. In 1965 this was declared to be a priority national project, and two years later, near Beijing, the first rocket engine test stand was constructed on the edge of a gully so that the flames could be deflected down the cliff face. The Long March engines underwent all-up testing there in June 1969. Afraid that this crucial task would be interrupted by the revolutionaries, Zhou Enlai issued orders that the tests were a matter of national honour, and that interfering with them would be regarded as unpatriotic. Thus protected, the workers toiled for eight days without interruption to complete the tests.

At Jiuquan launch site, construction of a new pad able to take the larger Long March 1 rocket and its new upper stage began in 1965 and was completed two years later. This was called 'area 2', and was close to the pad which had first fired the R-2 missile in 1960 (area 3).

Final testing of Long March and its upper stage took place during the course of 1968–69. The various parts were tested to exhaustion, the final rocket assembly being put together to test whether they would fit and function together (this is called the science of 'integration' in rocketry). The new rocket was successfully fired on 30 January 1970. This date is considered as the first successful launch of the Dong Feng 4 missile, or the Long March 1 without its third stage.

MOMENT OF TRUTH

All the different components of the Long March 1 were brought together for the first time in July 1969, even as the Moon race of the two superpowers reached its final, frenzied climax. A fully-integrated mockup left the works on 5 February 1970, and was shipped to Jiuquan, the real one following on 26 March. Tsien attended the launch preparations. Two versions of the satellite had been built, should a problem occur with one of them, and they arrived on 1 April. The next day, Zhou Enlai called a meeting in the Great Hall of the People to take progress reports on the preparations. The satellite was installed on the rocket, and a series of inspections and tests followed. Several days later, Tsien Hsue-shen, Ren Xinmin and their most senior colleagues reported to Zhou Enlai in Beijing that the satellite was ready for launch. The meeting continued late, but Zhou Enlai, eventually satisfied, approved the launch, sent the engineers back across China to the launch site and asked for daily reports following their arrival. Mao Zedong gave the go-ahead some days later.

The final assembly was erected on the pad on 17 April and cleared for flight on the 24th. The tower was moved back. That morning, the weather was warm and sunny with a spring breeze. The forecast that evening was fine: some clouds at 7,000 m and wind speeds of less than 4.5 m/s. Fuelling began. At 3.50 p.m. Zhou Enlai was on the telephone with a final good luck message. The day clouded over. 20 minutes before the launch time, 9.35 p.m., floodlights were switched on to bathe the gantry and rocket in milky light. Just then, the clouds parted, and stars began to wink through in the darkness of night.

At 9.35 p.m. that evening the historic 'Ignition!' command was given. Red and orange flames streaked out of the bottom of the Long March 1. China's first Earth satellite lifted off from its pad in Jiuquan and headed for orbit. Flames streaked across the night sky, at one stage making a tail 500 m long. At 60 s, the second stage began firing, even as the first stage was completing its burn. The first stage dropped off, to fall over Gansu. Telemetry signals were presented on coloured recorder pens in the control centre, marking the rocket's ascent to orbit. Each key stage was reported by loudspeaker, to be greeted with cheering. The second stage eventually tumbled into the South China Sea. The third stage, the capsule atop, glided upward. By the time the third stage ignited – 600 km high and 200 s later – the climbing rocket was over Guangzi. Everything went perfectly. At 9.48 p.m., launch control announced 'the satellite and rocket have separated and the satellite has entered orbit'. Only two minutes later, the anthem 'the east is red' was picked up. Tsien Hue-shen and his colleagues gathered on the still-hot launch pad – some cheering, some dancing, some even crying. Tsien made an impromptu speech. His life's dream had come true in the deserts of north-west China.

'WE DID IT THROUGH OUR OWN UNAIDED EFFORTS' – ZHOU ENLAI

China had become the fifth country to send a spacecraft into orbit. Strong signals were at once picked up by the American space command on 20,000 MHz. It was the heaviest first satellite launched by any country.

Zhou Enlai was telephoned with the good news at exactly 10 p.m. He said he would pass on the word to Chairman Mao Zedong at once, and that a celebration was in order. Later that night, he boarded a plane for a conference between China, Vietnam, Laos and Cambodia. He was able to tell them the news of China's achievement as soon as he arrived.

The official announcement was made the following morning. Zhou Enlai personally insisted that a small note be added to the press communiqué: 'We did it through our own unaided efforts'. The following evening, after the announcement had been made in Beijing, parades were held all over the cities, towns and villages of China. People vied with each other to be first to see the satellite in the spring night skies. In Beijing, fireworks were set off, bands played, coloured banners were unfurled. At 10.29 p.m. that night, Dong Fang Hong 1 passed over Beijing. Three nights later, it was spotted passing over Hong Kong, then a British colony.

Key steps toward a Chinese Earth satellite

January 1958	Proposal by Tsien Hsue-shen for an Earth satellite (project 581)
May 1958	Approval of project by Mao Zedong
January 1959	Project shelved indefinitely
1962	Recruitment of satellite team in Shanghai
January 1965	Proposal by Tsien to Central Committee for satellite launch
April 1965	Support from the party Defence, Science and Technical Commission
10 August 1965	Proposal approved by Prime Minister Zhou Enlai
May 1966	Design bureau 651 made responsible for satellite construction
24 April 1970	Launch of Dong Fang Hong 1.

Dong Fang Hong went into an orbit of 441 × 2,386 km, inclined at 68.55° to the equator and circling the Earth every 114.09 min. Though its silver–zinc batteries were designed to work for only 15 days, its transmitter continued to function for 28 days until the end of May. As hoped for, its signals could be picked up over great distances. Dong Fang Hong is still circling the globe. Air density will gradually drag it downwards, and it is expected to burn up in the atmosphere before 2070.

> **Press communiqué**
> Our great leader Chairman Mao has stated 'We too should produce man-made satellites'. In the midst of the triumphant march of the people throughout the country to hail the 1970s, we are happy to announce that this great call issued by China has come true. China successfully launched its first man-made satellite on 24 April 1970.

The launching was hailed in the Chinese media as one of the great events of the century. Granted the heightened level of political tension in China at the time, it is no surprise that the media announcements dwelled more on the politico-revolutionary portentousness of the event than its scientific import. Pictures of Dong Fang Hong were not released until ten years later. Chinese radio repeatedly rebroadcast 'the east is red' theme from the satellite for days. Details were given of when and where the satellite would pass over China. Photographs duly appeared of skywatchers scanning the heavens for a sight of the satellite (or, more likely, its larger and more visible rocket). China received an avalanche of congratulatory messages from all over the world – something it had not been used to for some time.

In fact, the satellite's main role appears to have been to broadcast 'the east is red'. Of each minute's broadcast, the first 20 s comprised 'the east is red'. This was followed by a

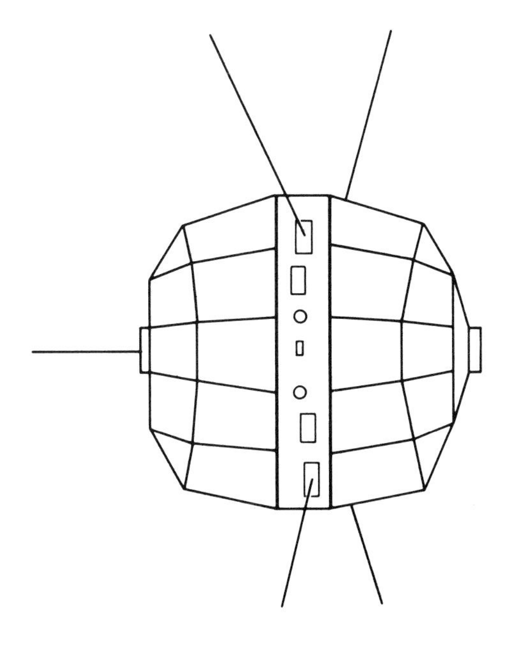

Dong Fang Hong

second rendition of 'the east is red'. After a 5 s gap, telemetry was transmitted for 10 s. After a further 5 s, the whole programme was repeated.

On the Mayday parade in Tiananmen Square a week later, Tsien Hsue-shen and Ren Xinmin stood on the podium along with other Chinese leaders as the band played the same tune that the first satellite was broadcasting all over the world. Mao Zedong commended Tsien Hsue-shen and Ren Xinmin in his Mayday address to the nation.

First satellites in orbit

October 1957	Soviet Union (Sputnik 1)	84 kg
January 1958	United States (Explorer 1)	14 kg
November 1965	France (Astérix)	40 kg
February 1970	Japan (Osumi)	38 kg
April 1970	*China (Dong Fang Hong 1)*	173 kg
October 1971	Britain (Prospero)	66 kg
July 1980	India (Rohini 1B)	35 kg
September 1988	Israel (Offeq 1)	156 kg

DONG FANG HONG AFTERMATH

In August 1970, after the Dong Fang Hong launch, the Lin Biao revolutionaries were dominant in the Military Commission which directed much of China's planning. The Military Commission persuaded the government and party to adopt a new five-year plan (1971–76) which had the slogan 'three years catching up, two years overtaking', committing the country to developing eight new launch vehicles and 14 new satellites in five years (other reports speak of an average of nine satellites a year). Many of these projects, which most scientists considered to be unnecessary and unrealistic, got under way, though few saw the light of day. This was disruptive to existing projects and saw the commencement of projects which later had to be abandoned.

The very survival of the space programme in this time of turmoil may in many ways be attributed to Zhou Enlai. He went to considerable efforts and probably some political risks to put the space industry off-limits to the revolutionaries, and he also drew up lists of the scientists involved in the Earth satellite project, putting them under protection from vilification. When Zhou Enlai died, personnel in the space industry turned out in vast numbers to mourn him, placing wreaths and poems in his honour in Tiananmen Square, despite large scale intimidation by the Red Guards.

SECOND SATELLITE: SHI JIAN 1

China's second satellite followed Dong Fang Hong into orbit nearly a year later. Now that the propaganda value of launching a first satellite had been demonstrated, China's second satellite could concentrate on scientific tasks. It is possible that this second satellite achieved what the first one had been intended to do, had not political and propaganda imperatives risen to the fore in 1966, during the early design stage.

The tasks of the second satellite were agreed at a conference held in the Chinese Academy of Space Technology in Beijing during May 1970. The second satellite received a new designation: Shi Jian ('practice'). Sun Jiadong was in charge of the overall project. The first mockup was assembled in 1969, and full testing was carried out on three engineering models throughout 1970. There were delays in May 1970 when staff in the Shanghai Scientific Instruments Plant were reassigned by the Red Guards to work in the countryside. Some scientists were put under political investigation and others were put under house arrest; and they were not helped by Lin Biao, who in October 1970 dispersed the Space Physics Institute from Beijing.

Slightly heavier than Dong Fang Hong at 221 kg, Shi Jian eventually entered orbit on the evening of 3 March 1971. Signals were transmitted on 20,009 MHz and 19,995 MHz. Although the explosive bolts separating the satellite from the third stage had fired, the satellite did not separate from its carrier rocket. Enveloped within the third stage, the signals received were weak – only about 1 per cent of what had been hoped for. The designers were, as one might imagine, perplexed and worried. On the eighth day, the signals suddenly came through loud and clear. Ground observations confirmed that the satellite had now separated from the launcher. The reasons for the failure to separate, and the sudden and unexpected release, were never understood.

The satellite's orbit was similar to Dong Fang Hong – 267 × 1,830 km, inclination 69.9°, period 106.15 mins. Beijing did not announce the launch until 16 March, presumably when separation had been confirmed and stronger signals had been received.

Shi Jian appears to have been a modified version of Dong Fang Hong – a 72-sided polyhedron. The chemical battery system was taken out and replaced by a nickel-cadmium system rechargeable by solar cells. In place of the tape recorder which played 'the east is red', Shi Jian carried three scientific instruments – an 11 mm cosmic ray detector, a 3 mm X-ray detector and a magnetometer – and a heat flow meter was also carried. A hundred automatic thermal shutters closed as the spacecraft entered darkness, opening again as it entered light (a similar system was carried by Sputnik 3).

Using four short-wave antennas, the radio transmitter emitted a stream of scientific data on 16 channels (this time there were neither periods of silence nor music) and could be picked up 3,000 km away. The system's power needs were under 2 W. Shi Jian continued to transmit scientific data until it burned up in the upper atmosphere on 17 June 1979. The battery and telemetry systems showed no evidence of deterioration and maintained the same high level of performance throughout the mission, despite 10,000 charging and recharging cycles (one for each orbit as the satellite went into and came out of darkness). The solar louvres, likewise, opened and closed automatically a similar number of times. The design teams rightly received commendations for these achievements in 1978. The 3,028-day mission appears to have been completely successful.

EPILOGUE: TSIEN HSUE-SHEN

By this stage, Tsien Hsue-shen's ambitions of the 1940s and 1950s of launching a satellite had been fulfilled. Tsien must rate as one of China's greatest scientists of the twentieth century. Now in his late 80s, he does not grant interviews, but he gave permission to his secretary to write a biography, with the condition that work on it must not commence until after his death. Although he made a brief visit to the United States in November 1972, he refuses to return there, even to receive academic awards. He is still angry about the misery he endured there in the 1950s and the failure of the government of the United States to apologise for its wrong-doing.

Chang's view is that Tsien made four main contributions to the resurrection of Chinese science in the 1950s. First, he gave the Chinese leadership, Mao Zedong and Zhou Enlai, the confidence that by investing in rocketry the money would be well spent and there would be positive outcomes. Second, Tsien was able to bring discipline and coherence to the engineers and scientists charged with modernising China's rocket forces. Third, he helped to build up the intellectual infrastructure for the development of Chinese science. In Russia at this time, scientists were obliged to build up a totally indigenous space industry, though in practice a very small group had secret access to western materials[12]. Tsien insisted that his scientists and engineers build up a proper system of reference books and materials, not just in Chinese, but in Russian and English. Fourth, Tsien built up the organisation necessary to ensure that China developed a proper space programme. He established China's Institute for Missile Design and all the other important elements that are essential for a national, coordinated space effort.

Although deported from the United States in 1955 for being a communist (which he almost certainly was not), he joined the Communist Party of China, in China, the following year, and from then on followed the party line, even issuing statements to denounce colleagues who stepped out of line, and supported the great leap forward.

In late 1975, Tsien made the mistake of attacking his superior, Zhang Aiping, accusing him of national chauvinism in a poster which was widely circulated at the time. Zhang Aiping was then trying to sort out the problems around the failed first launch of the Long March 2 rocket. Zhang Aiping was not just anybody, but had been a senior revolutionary and military leader in the 1960s and 1970s. He was then Minister of the Commission of Science and Technology for the National Defence. The Gang of Four led the chorus in attacking not only Zhang Aiping but also Deng Xiaoping, then chairman of the military committee of the chiefs of staff of the People's Liberation Army. Tsien accused Deng of being 'the sworn enemy of all scientific workers who take the revolutionary road'.

Tsien was now publicly associated with the Gang of Four, who ruled China following the death of Mao Zedong in September 1976. Their ascendancy was short-lived. They were deposed in a military coup the following month and were sent to cool off in prison. Deng Xiaoping became, by early 1978, chairman of the communist party, a position he held until his death in 1997. Tsien's old boss, Zhang Aiping, was reinstalled.

Unsurprisingly, Tsien became increasingly marginalised in China's space leadership. He tried to recant by writing newspaper articles explaining how it was really the Gang of Four who had retarded China's scientific progress. He managed to get back on side in 1989 when he publicly rushed to support the action of the leadership of Deng Xiaoping and Li Peng in their bloody suppression of the demonstration in Tiananmen Square. In 1991, the government bestowed on the 80-year-old the award of State Scientist of Outstanding Contribution.

Iris Chang's biography of Tsien Hsue-shen concludes that Tsien was not an original rocket theoretician. Had he died in 1955 or never gone to China he would have merited only a footnote in the history of science. His return to China was the crucial moment, for he was able to apply his American knowledge to China's growing military ambition. His real skills were in leadership.

ASSESSMENT AND CONCLUSIONS

This concludes the account of the first stage of the Chinese space programme. The return of Tsien Hsue-shen to China coincided with a period of military and scientific expansion. Concepts of Earth satellites were first explored in China in 1956, just as designs for early Earth satellites were being hardened up in the United States and the Soviet Union. The prospects that China might be the third space power, or enter the space race not far behind the Soviet Union and the United States, were dashed because of the political upheavals of the great leap forward (1958–59) and then the Cultural Revolution (1966–76).

Instead, efforts by China to build an Earth satellite were peristaltic, with periods of rapid progress coming to nothing because of political upheaval, the inability of the political leadership to make or execute decisions and endemic issues about who should lead the Earth satellite programme, leading to many complicated organisational changes. As a result, China became a space power much later than it could have been. Even then, the first satellite, sent up in the course of the political turbulence, was used for political purposes. Not until the following year, 1971, did China put a proper scientific satellite into orbit. Order eventually prevailed over chaos. Shi Jian 1, while little publicised and hardly remembered compared with Dong Fang Hong, was an outstanding success, returning scientific data from three instruments throughout the 1970s – a vindication of the engineers who had designed and built the spacecraft.

3

Expansion of the Chinese space programme (1972–84)

Developing new confidence from the launch of Dong Fang Hong and Shi Jian 1, the 1970s were to see a considerable expansion of the Chinese space programme. The decade saw the programme recover spacecraft from Earth orbit, build two new launchers and put three scientific satellites into orbit. The main line of the development was the recoverable satellite programme, in which three satellites had been launched by the end of the decade and nine by the end of the first series. In operating a recoverable satellite programme, China became the third country, after the United States and Soviet Union, to bring satellites back from Earth orbit. In addition to these achievements, China put into orbit three mysterious satellites in the Ji Shu Shiyan Weixing series. Over 20 years later, their purpose has never been explained.

'NO SPACE RACE'

Like the 1960s, the 1970s were affected by political turmoil, though it was ultimately less destructive than the first phase of the Cultural Revolution. There were fresh political interruptions after the dramatic events of September 1971 when Lin Biao, one of Mao Zedong's most trusted lieutenants, fled China for the Soviet Union: en route his plane was shot down by Chinese fighters and it crashed in flames. The consequent purges affected the Seventh Ministry, its head, Wang Bingzhang, being jailed for many years. Paranoia was rife. Builders of a cosmonaut training device were implicated, accused of preparing equipment to cure Lin Biao's insomnia. In 1976 there were further purges of those scientists thought to be sympathetic to Den Xiaoping, who had been sacked by Mao Zedong. Order did not return until after the death of Mao in September 1976 and the overthrow by the military of the Gang of Four the following month. Once this took place, a rectification campaign was organised. This restored discipline in the space industry, and reinstated rank to scientists who had been unjustly accused by the revolutionaries. Pre-Cultural Revolution working procedures were put back in place.

Key events in Chinese political history

1949	Chinese revolution
1958	Great leap forward
1966	Cultural Revolution
1976	Ascendancy of Gang of Four
	Death of Mao Zedong
	Jailing of Gang of Four
1977	Ascendancy of Deng Xiaoping
1978	The four modernisations; the period of rectification
1989	Crushing of democracy movement
1997	Death of Deng Xiaoping

The 1971–6 plan, which had involved (at least on paper) a furious expansion of the space programme, was scrapped, and a new, more realistic plan was drawn up by Zhang Aiping. He defined the key tasks for the space programme in the 1980s as being to put Dong Feng 5 into operation as an intercontinental ballistic missile, to launch a geostationary communications satellite, and to develop a submarine-launched missile. This less ambitious but more definite and practical plan was promptly approved by the Central Committee. The new leadership under Hua Guofeng and Den Xiaoping encouraged newer, younger and more pragmatic engineers and managers to come forward in industry, concentrating on modernisation rather than ideological struggle. At the same time, it was decided to reduce defence spending from 12 per cent to 5 per cent of national income, demobilise a million soldiers, and convert some military facilities to civilian use.

In August 1978, Deng Xiaoping articulated what he believed China's space policy should be. Receiving a report from the Seventh Ministry, he told them that China was a developing country: 'As far as space technology is concerned, we are not taking part in the space race. There is no need for us to go to the Moon and we should concentrate our resources on urgently needed and functional practical satellites.'[13]. The space budget was trimmed to meet its new, more modest ambitions and fell to 0.035 per cent of Gross National Product, trailing not only the big space powers but also Japan (0.04 per cent) and India (0.14 per cent).

FOUR MODERNISATIONS

In October 1978, Deng Xiaoping announced the 'four modernisations' for China in the post-Mao epoch: science and military technology, agriculture, education and industry (dissidents added the fifth modernisation, that of democracy). Hand-in-hand with the four modernisations went an opening up of the economy and science. Foreign investment was welcomed, significant areas of the economy (for example retailing and services) were privatised, special economic zones were identified and developed, and cooperation in technology was promoted. The four modernisations were approved by the 3rd plenary session of the 11th party Central Committee in December 1978. In a pointed reference to the space industry, it reaffirmed that quality control was the chief task of scientific production, rather than the intensification of the class struggle.

However, the effects of the Cultural Revolution could not simply be undone by party resolution. China's technical schools and universities had been effectively closed for ten years. Virtually no new graduates had come into the space industry, and there was an acute shortage of scientific labour.

One of the features of rectification was the restoration of the chief designer system. This system, introduced in China in May 1964, was Soviet-originated (*glavnykonstruktor*, in Russian; the first Soviet chief designer appointments dated to 1946). In China, each chief designer was assisted by a technical designer and an administrative designer. Each chief designer had a general department in his bureau which was responsible for leading and coordinating the project in question. The following chief designers were appointed:

Tu Shoue	DF-5 missile
Huang Weilu	Submarine-launched missile
Ren Xinmin	Satellite communications project
Sun Jiadong	Dong Fang Hong 2 comsat
Xie Gyuangxuan	Long March 3 rocket

Openness and purposeful modernisation, with science at the top of the list, provided a much more promising environment in which a space programme might flourish. In 1980, China joined the International Astronautical Federation, though on condition that it displace nationalist Taiwan. The Chinese membership body was the Chinese Society of Astronautics, established the previous year with Tsien Hsue-shen as its honorary president and Ren Xinmin as its president. Forty scientists subsequently became academicians, with Ren Xinmin becoming a member of the board of directors. In 1983, Yang Jiachi was elected executive committee vice-president. Tu Shoue was made chairman of the educational commission. China also joined the International Telecommunications Union and the UN committee on the peaceful uses of outer space.

In effect, China's 20-year isolation from the world space community was quickly brought to an end. In 1977, Chinese space experts visited France and, the following year, Japan. In 1979, China received visits from the European Space Agency, France, Japan and the United States. An American delegation went to China. The first of many Chinese regional and international conferences on space were held in 1985. In due course of time, the Xi Chang space centre even came to feature on tourist itineraries.

While awaiting the development of its own Earth resources satellites, China built its own ground station near Beijing to receive American Landsat data in 1986. In 1988, China sent its most promising engineering graduates to courses at the Massachusetts Institute of Technology – the first time they had studied there since their predecessors had been driven out during the 1950s.

The Chinese space programme opened up within China itself. Workers in the space industry had been prohibited, on pain of extreme penalties, from telling their families where they worked. In a practice borrowed from the Soviet Union, they were assigned to mail box numbers, their institutes never being geographically identified. The greatest challenge faced by new graduates assigned to the space industry was to find their future place of work, since virtually no one was allowed to tell them where it was; likewise, the railway line from Qingshui to Jiuquan was not to be found on any map. The former policy not only concealed the space programme from foreigners and frustrated graduates, but also inhibited cooperation between scientists within China itself. From now on, space organisations were publicly named, identified and listed.

PROJECT 701: THE JI SHU SHIYAN WEIXING SERIES (1973–76)

There was a gap of more than four years between the launch of Shi Jian 1 (1971) and the next Chinese satellite (1975). In the event, the next series of satellites, produced before the period of openness and modernisations, raised more questions than it answered. The series comprised three successful launches and one failure during the period 1973–76. The series has been mentioned but never described in the Chinese literature. In China, it was code-named project 701. Construction of the Ji Shu Shiyan Weixing satellite had begun in January 1970, although virtually nothing is known of its development or history.

Ji Shu Shiyan Weixing translates as 'technical experimental satellite'. Because so little information has been made available about the programme, it is assumed that it may be military. It may have been an attempt to develop a satellite for electronic intelligence gathering, which has been a dominant theme in the military satellite programmes of the Soviet Union and the United States. It could also have been photographic in purpose, though it is unclear how the images were returned. The series took place at the same time as the development of the Chinese recoverable satellite programme, and in the absence of information from China about either programme, they were several times confused. When the first launching took place, the official announcement appeared to confirm the military thrust of the programme, stating that the satellite was part of 'preparations for war'. The subsequent official history refers to the importance of the satellite entering a very precise orbit and that small errors in perigee were simply not acceptable[14]. This was to be a familiar characteristic of Soviet electronic ocean intelligence satellites, so it is possible the Chinese series had a similar purpose.

THE FENG BAO ROCKET

Project 701 used a new launcher, the Feng Bao, made in Shanghai. The chief designer was Shi Jinmiao. Feng Bao was based loosely on the Dong Feng 5 missile, data from which were passed to the design team in Shanghai. Responsibility for the Feng Bao was given to the Shanghai #2 Bureau of Machinery and Electrical Equipment, and was built in the Xinzhong Hua plant. The subsequent official histories of the Chinese space programme looked askance at this decision:

> The Shanghai region had never previously undertaken the research and development of a carrier rocket and was ill-equipped to do so in every respect, be it technological resources, production capability or testing facilities to research and develop a large rocket [15].

There appear to be two reasons for the decision to build the new rocket in Shanghai. One was probably political; it was Mao Zedong's power base and he probably liked to allocate pet projects there. (This was not just an eastern habit; Lyndon Johnson, after all, moved mission control from Cape Canaveral to his home state of Texas, 1,500 km distant). The other reason is that there may have been an intention to build up a space industry base outside Beijing as well as in the national capital. Shanghai was the most advanced industrial city in the country and was the best candidate. The bureau received considerable help from the Chinese Academy of Launcher Technology in Beijing, which received study teams and sent them specialised personnel and the blueprints of Dong Feng 5.

SHANGHAI MOBILISES

The Shanghai team was quick to mobilise to the full the industrial and technological resources of the city and the region, quickly drawing in all the leading state and municipal companies. A local computer company, the Hua Dong Computer Technology Research Institute, designed the on-board computer, apparently from scratch. The rocket's aluminium–copper alloy tanks were welded by the Shanghai Jiangnan shipyard. These team members were very resourceful: when the local factory was just too small to carry out necessary welding work for the Feng Bao, they jacked up the roof of the building by 1.7 m to gain the extra space.

Unlike Beijing, the Shanghai region had no testing equipment. Undeterred, the team managed to adapt existing factories for their purposes. Within a few months, they had built a final assembly building, shake table, engine test stand and materials strength-testing unit. Despite their inexperience, the Feng Bao team managed to build and evaluate the test version of the rocket in only ten months. This was no mean achievement, for its performance was somewhere between that of Long March 1 and Long March 2, where equivalent work had taken the designers four years. The Beijing team, whilst annoyed at scarce resources being used elsewhere to match their efforts, came to admire the speed and talent of the Shanghai designers.

FIRST HOT TESTS

A non-flight test version of the Feng Bao left for Jiuquan in November 1970, just over a year after the team had started work. In March–April 1971, the engines were hot-tested in Jiuquan and worked perfectly. The test revealed a number of problems, such as the computer and poor valves for the first stage engine, all the result of rushed work and lack of rigorous quality control. After a campaign of improvement and renewed quality control, the first flight test rocket was completed exactly a year later when it was sent by train to Jiuquan in April 1972. Not until 6 August did the design team feel able to tell Zhou Enlai that the rocket was ready.

The first test of Feng Bao – a sub-orbital mission – took place on 10 August 1972. Although a success in some respects, the attempt revealed areas where further improvement was required in order to carry a heavier payload. Fuel tanks were recast with thinner walls, fuel flow to the engines was improved and it was decided to run all engines to exhaustion in the ascent to orbit. The manoeuvring engines would be used to put the rocket into orbit. Cables were re-wired, redundant equipment removed, and miniaturised systems introduced. As a result of these measures, the payload increased by 50 per cent. Some of the equipment was tested against leaks for as long as a year at a time; other equipment was alternately baked or cold-soaked for weeks on end.

Chinese engineers seem to have encountered considerable difficulty with the Feng Bao design. The rocket was substantially more demanding than Long March 1, being required to lift payloads of 1.9 tonnes compared with the 300 kg of the first Long March. Official histories have blamed the difficult history of Feng Bao on the interference of the Cultural Revolution, with its negative effects on quality control during the crucial phases of design[16]. The revolutionaries had an especially strong following in Shanghai, where the Feng Bao was built. Scientists who tried to confront technical problems were accused of sabotage and acting on behalf of class enemies.

IN ORBIT AT LAST

The engineers' best efforts were to no avail, for when the first Feng Bao was launched again on 14 July 1974, thrust in the second stage vernier manoeuvring engine collapsed and the satellite failed to reach orbit. In later accounts, the Gang of Four was blamed, but the technical decision was taken to replace the second-stage engine with engines taken from the already proven Long March 1. Eventually, patience was rewarded on the second attempt. On 26 July 1975, the Ji Shu Shiyan Weixing 1 entered an orbit 183 × 460 km, 69°, 91 min. The launch announcement gave the barest details about the satellite (e.g. orbital parameters) but a weighty political commentary (the final struggle around the succession to Chairman Mao was under way). Ji Shu Shiyan Weixing 1 decayed after 50 days in orbit, crashing into the atmosphere over the Pacific Ocean on 14 September 1975.

Ji Shu Shiyan Weixing 2 entered orbit on 16 December 1975. This time, the launch announcement did not even given the orbital parameters, instead putting it in its more appropriate context of the movement to criticise Lin Biao and Confucianism. Ji Shu Shiyan Weixing 2 entered a slightly lower orbit than Ji Shu Shiyan Weixing 1 (183 × 387 km, 69°, 90.2 min), burning up in the atmosphere after only 42 days.

Ji Shu Shiyan Weixing 3 came nine months later, on 30 August 1976. It had markedly different orbital characteristics, flying much further out than its predecessors (198 × 2,100 km, 69.2°). Like its two predecessors, it weighed 1,110 kg. The launch announcement gave even fewer details (only the date) about the satellite, paying greater attention to its political significance (it marked the struggle against the right deviationists). The satellite decayed in 817 days. None of the three satellites manoeuvred in orbit; nor were signals picked up in the West, and presumably they were transmitted only over China itself.

The programme then closed, which may have been because it did not achieve the intended results. Officially, they were technology test satellites, but it is not clear what technology was tested nor how it was subsequently applied. Another explanation lies in the fact that both the satellites and the launcher were built in Shanghai, which was Chairman Mao Zedong's political base and that of the Gang of Four. The scientists and engineers there may have lost support when Mao died in September 1976. The failures of Feng Bao were retrospectively blamed on the Gang of Four, and the engineers were reassigned to work on Long March 3. American aerospace experts visiting China in 1979 were shown what they were told was a spare military spacecraft in the Shanghai Huayin Machinery

Plant: a cylinder 2.5 m tall, 1.7 m diameter, weighing 1.2 tonnes, with 1 cm × 2 cm solar cells. But they were none the wiser as to its precise purpose. The official history, many years later, makes reference to costly projects hastily entered into without proper discussion during the period of the Cultural Revolution[17]. This could be an oblique reference to this programme.

Table 3.1. Ji Shu Shiyan Weixing series (1973–76)

Name	Launch date	Notes
Ji Shu Shiyan Weixing 1	26 July 1975	Decayed after 50 days.
Ji Shu Shiyan Weixing 2	16 December 1975	Decayed after 42 days.
Ji Shu Shiyan Weixing 3	30 August 1976	Decayed after 817 days.

THE RECOVERABLE SATELLITE PROGRAMME (FANHUI SHI WEIXING): PROJECT 911[18]

China was the third country to recover a satellite from Earth orbit. The idea of a recoverable Earth satellite in China goes back to 1964 and the work of the Shanghai design team. They had been inspired by what they had read of the American Discoverer series of recoverable satellites in the late 1950s and early 1960s; indeed the eventual design of the return capsule bears clear similarities with Discoverer. Whether they realised that the Discoverer programme was a military programme (it was declassified in 1995 and revealed as the Corona programme), designed to film Soviet missile bases and return a film-bearing capsule to Earth, is not known.

The idea of a recoverable Chinese satellite was first formally proposed in the Chinese Academy of Sciences *Proposal on plan and programme of development work of our artificial satellites* (spring 1965). The Shanghai team applied for responsibility for this task, which was awarded to them in spring 1966. Conceptual studies were carried out by Wang Xiji throughout 1966, and detailed design work was carried out the following year. Wang Xiji's team settled on a satellite weight of 1,800 kg, with a typical orbit of 173 × 493 km, 59.5°, 91 min. A three-day conference on the progress of the project was held from 11–13 September 1967, and the programme was given a code (project 911). It was given the name Fanhui Shi Weixing (FSW) – 'recoverable experimental satellite'.

The precise purpose of the programme has never been entirely clear. If it was modelled on the Discoverer programme, as seems likely, it was probably photoreconnaissance. However, later versions of the Fanhui Shi Weixing were also used to conduct a range of microgravity experiments in orbit. The cameras were put to civilian use. Whether this was because of an improvement in the international climate, or the limited reconnaissance value achieved from the Fanhui Shi Weixing missions, or a form of diversification, is a matter of guesswork. The military intelligence benefit derived from orbiting a recoverable cabin for a week every year is probably very limited.

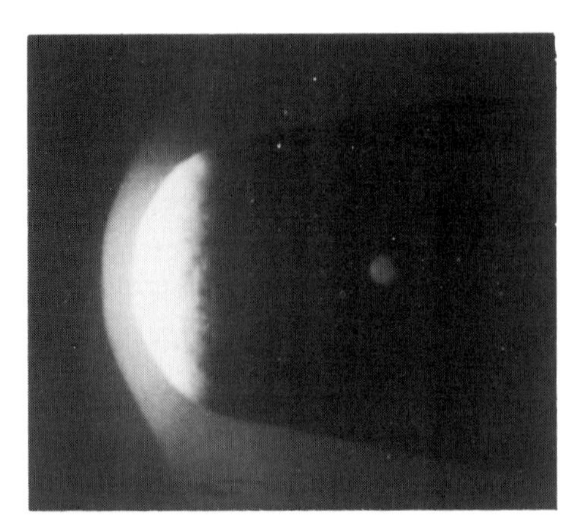

Heat tests of FSW.

The task of developing the project fell to the Chinese Academy of Space Technology (CAST) which by the early 1970s had grown to 81 units spread over 16 different provinces. A new rocket, Long March 2, was developed by the Chinese Academy of Launcher Technology (CALT). Progress was held up by the later phases of the Cultural Revolution, but for which the first launch might have occurred several years earlier than it eventually did.

THE CHALLENGES

Recovering satellites poses several difficult engineering challenges: devising a protective heat shield to ensure the capsule survives re-entry temperatures of 1,200° C, the development of retro rockets, a very precise attitude control system, quality ground tracking to prepare the cabin for the precise moment of re-entry, and search and recovery systems. These engineering challenges required the development of ever more sophisticated ground testing equipment. In 1970, a thermal vacuum chamber called the KM_3 was constructed by the Institute of Environment Test Engineering and the Lanzhou Institute of Physics. The level of vacuum achieved was 10^{-9} torr (1 torr = 1/760 atmospheres). A tracking centre was built: the Xian Satellite Surveying and Control Centre.

The Chinese had no previous experience of making heat shields. They did not wish to use ablative heat shields of the type developed by the United States and Soviet Union in the 1960s: these were heavy shields, in which the material progressively burned off during the descent, enough remaining for the cabin to survive. Equally, they knew they did not have the capacity to go straight to low-density foam-type shielding, of the type subsequently used by the American and Soviet shuttles (tiling). They eventually found a non-ablative material whose qualities lay somewhere in between: a carbon composite material called XF, able to withstand re-entry temperatures of 2,000° C. The main aerodynamics and ablation expert involved in the re-entry problem was Zhuang Fenggan. Like Tsien, a graduate of Jiatong University, he graduated from the University of California College of Science in 1950. He was responsible for the construction of wind tunnels to test the spacecraft.

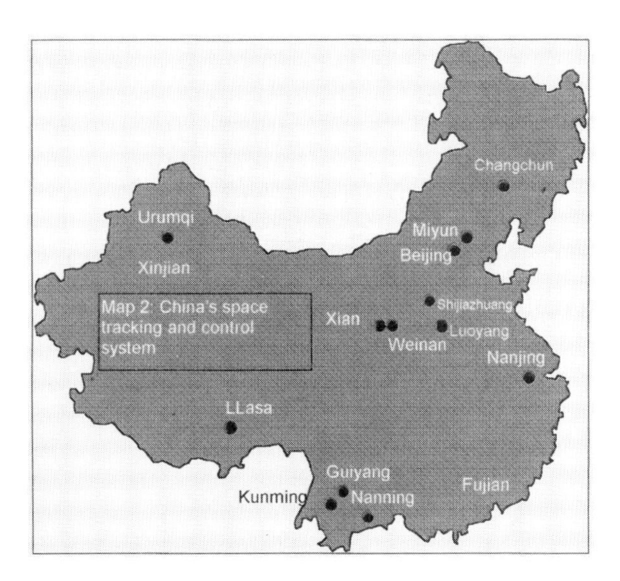

Map of China's space tracking and control system.

The recoverable series required a relatively advanced level of automation. The person responsible for the automation in Chinese satellites was Yang Jiachi (b. 1920), also like Tsien a graduate of Jiatong University. He received his doctorate at Harvard, where he worked on early computers. After his return to China in 1956, he concentrated on automatic control systems in space. A new three-axis attitude control system was developed by Zhang Goufo of the Beijing Control Engineering Research Institute, using an infrared horizon scanner and a gyrocompass. The scanner was tested on two T-7A sounding rocket tests at Jiuquan in June and July 1969.

A camera system was developed by the Changchun Institute of Optics and Fine Systems, which had some previous experience of aerial cameras but none of space or stellar imaging (more sophisticated charge coupled device (CCD) cameras were introduced later in the series). The early series carried cameras for ground photography and side-pointing cameras for stellar photography (so as to work out the precise attitude of the spacecraft in orbit). The man credited with the design of the ground imaging was Wang Daheng, an optics specialist whose previous work involved the design of rocket and satellite tracking devices. The designer of the stellar camera was Wang Jintang. They had to start from scratch and trawl foreign and domestic literature on advanced cameras as their starting point, but in the midst of their efforts the turmoil of the Cultural Revolution forced them to leave Changchun and move operations to Beijing. They produced the first cameras in 1969, testing them on aircraft first, and having to overcome a series of problems such as films jamming.

One of the most crucial links in the chain was the retro-rocket, which would be required to fire the Fanhui Shi Weixing out of orbit to begin the final descent. The solid-fuel rocket would be required to fire in a vacuum after orbiting the Earth for several days in weightlessness. Ten tests were carried out during 1971–1974, in the course of which the problems associated with the new engine were resolved.

The parachute system proved problematical. The importance of a safe system was underlined in April 1967 when Soviet cosmonaut Vladimir Komarov died when his parachutes snarled during an already-troublesome return to Earth. In July 1970, the Chinese air-dropped two FSW capsules to test the parachute system. Both failed and the cabins were destroyed. A second set of tests in October had an identical dismal outcome. With the help of the Air Force, drop tests went on, and by 1974 they were confident that a reliable system had been devised.

The Fanhui Shi Weixing satellites were a quantum leap in size and scale beyond the first two satellites. They were specified to weigh in the order of 1,800 kg – the actual payload being in the order of 150 kg – and involve both a service module and a recoverable capsule. The cabin itself was beehive-shaped, being 3.1 m tall and ranging from 1.4 m in diameter at the forward end to 2.25 m in diameter at the large end. The normal procedure was for the 800-kg service module to remain in orbit for about 20 days.

The conceptual studies made in the course of 1966–67 were interrupted by the Cultural Revolution. When they resumed in the late 1960s, prototypes were manufactured, engineering and vacuum tests were carried out to verify the limits of the cabin's abilities, and recovery teams practiced finding returning capsules. Perhaps the last item to be completed was the solid-fuel retro-rocket, the design of which was finally qualified for flight in 1974.

Leaders of the Chinese recoverable satellite programme (project 911)

Zhu Yilin
Wang Xiji
Sun Jiadong

MISSION PROFILE

The Fanhui Shi Weixing satellites consist of a blunt cone capsule placed on a service module. During the mission, the nose is pointed in the direction of travel. At the end of its mission, the spacecraft is swivelled through 100°, pointed directly toward the Earth, and the retro-rockets are fired. This is a crude method of returning to Earth as it uses up a substantial amount of fuel, but it has the advantage of ensuring that retrofire can be commanded and take place over China. By contrast, Russian spacecraft returning to Earth make a more gentle and economical braking manoeuvre in the direction of travel over the South Atlantic, the spacecraft descending in an arc that brings them high over the Mediterranean.

The angle of retrofire must be very accurate, as each degree of error results in a 300-km difference in the landing site. At 16 km, the FSW drops its heat shield and retro-rockets, a parachute opens and the cabin descends at 14 m/s over Sichuan province in southern China. The Chinese approach to re-entry, with retrofire over China, has the advantage of making sure that the cabin returns to Chinese soil, an important factor if sensitive film is on board. (The Russians fitted self-destruct devices to their spacecraft to stop them falling into unfriendly hands). On the other hand, the Chinese re-entry manoeuvre requires a velocity change of 650 m/s – much more than the standard Russian or American re-entry profiles (about 175 m/s).

Sichuan province, in the south-west of the country, was chosen as the recovery zone, although it was hilly and often subject to thick clouds and mists. Photographs from the recovery area have always shown Mil-type recovery helicopters hovering against a background of mountains, following the descent craft down and then lifting it away for post-flight examination. The scene is one of the space cabin lying on the hillside – its red and white parachute streamed out alongside – the recovery teams safing and checking the cabin, and rural workers gathering on the nearby hills to watch the excitement.

LAUNCHER FOR THE FANHUI SHI WEIXING: LONG MARCH 2

The much greater weight of the payload would require more lifting capacity. Accordingly, Long March 2 was developed by the Chinese Academy of Launcher Technology (CALT) in Beijing. CALT used as its starting point the design of the Dong Feng 5 ballistic missile. The chief designer of Long March 2 was Tu Shoue. The new rocket had two stages, was 32 m tall, used nitrogen tetroxide as oxidiser and unsymmetrical dimethyl hydrazine (UDMH) as fuel, weighed 190 tonnes, and had a liftoff thrust of 280 tonnes. Made of high-strength aluminium copper alloy, it was the first Chinese launcher to use full computer guidance and gimbaled engines.

Long March 2 en route to launch site.

The initial design of Long March 2 took four years, from 1965 to 1969. The payload requirement of almost 2 tonnes was substantially greater than the requirement for Long March 1, then also in design but as yet unproven. The main technological challenges were to build an inertial guidance platform to guide the rocket during its ascent (essential for placing the satellite in the precisely required orbit), the development of an on-board computer (essential for accurate recoveries) and the use of lighter metals (aluminium–copper). In reality, the main challenges were political – progressing with the work despite the interruptions of the Cultural Revolution.

New engines were designed for Long March 2. Called the YF-20, four were used together on the first stage (when so clustered, they were called the YF-21). A version of the

YF-20 was used for the second stage (the YF-22). Li Boyong led the design of the new engine, a process which involved many heartbreaks, with recurrent problems of vibration and turbopump failures. The four engines were first tested all together on 14 June 1969. Crowds gathered on the hills adjacent to the test stand. They then cheered to the roar of the engines, whose thunderous rumble was matched by the hissing of 7.9 tonnes of water per second issued from 30,000 nozzles in an attempt to keep the test stand cool. The second stage was given a similar test in November 1970.

A key innovation with Long March 2 was that the engines could be swivelled about their axis to steer the rocket (in rocket science, this is called gimballing). Hitherto, Chinese rockets had used vanes, or fins, for steering, a technique introduced by the Germans on the A-4. Gimballing saves the weight of vanes, and is more precise, but requires a much more sophisticated engine design. To steer the second stage, a complex of four small vernier engines (the YF-24) was used. The first gimballing tests were made on a test stand in December 1969. The second-stage engine was tested out at the rocket engine testing station in late 1970. The engines required for the Long March 2 represented a considerable jump in performance, from 28 tonnes to 73 tonnes of thrust.

A further innovation was the use of self-pressurisation: instead of the oxidiser and fuel tanks being pressurised by separate tanks of pressurising gas, a vaporiser was used, effectively cutting out the need for separate gas bottles. The propellant tanks were made of aluminium alloy, producing a weight reduction of 30 per cent on the aluminium–magnesium alloy of Long March 1, though the material posed formidable welding problems.

Computerised guidance systems for Long March 2 were designed by Liang Sili and the Micro-Electronics Research Institute. The lightweight medium-speed, small-capacity digital computer was the first of its kind in China. The Chinese found the development of advanced guidance and computer systems to be a problem. China lagged far behind the West, and the Soviets had been of little help in the 1950s. (In Stalin's time, computers had been condemned as a 'false bourgeois science', and the Russians had tried desperately to keep up ever since). Two Chinese scientists were called in to resolve these problems: mathematics experts Liang Sili and Zheng Yuanxi of the Inertial Devices Research Institute. The computer for the CZ-2 was designed by the Micro-Electronics Research Institute, established only in 1965. Significant improvements in space electronics at the time may be attributed to Wang Zheng (1890–1978), veteran of Mao's army in the 1930s, and the man who modernised telecommunications, computers and electronics in China.

Another innovation, developed by designer Yu Menglun, was the interstage glide. Once the second-stage engine had completed its burn, the manoeuvring vernier engines would continue to fire as a main engine. Also called the low-thrust orbital entry technique, this approach enabled an extra 500 kg of payload to be carried, though it meant that the ascent to orbit took several more minutes.

Having done all this, the designers then organised a campaign to reduce the weight of the rocket still further, each kilo saved meaning more payload. By refining the equipment still further and by using lighter alloys, they achieved savings of 700 kg.

The quantum leap forward of Long March 2, compared with Long March 1, even before the latter had flown successfully, required that testing procedures should be more exacting than ever. The entire assembly was tested in the vibration testing tower and shaken time after time against the various vibration loads that affect rockets from side to

Closing out Long March 2 for launch.

side and end to end. After these tests, the rocket would be stripped down and every part examined to determine the effects of stress.

Premier Zhou Enlai inspected the Long March 2 in September 1972. It was one of 30 occasions on which he had met with scientists, engineers and technicians involved in the rocket programme. Encouraging them in their efforts, he reminded them of the absolute importance of quality control, safety and reliability, and of not being blown off course by the cultural revolutionaries (or 'left deviationists' as they were termed pejoratively at the time).

Designers of Long March 2 (CZ-2)

Tu Shoue
Wang Yongzhi
Wang Dechen
Li Zhankui
Yu Menglun

BROKEN WIRE

The first attempt to launch a recoverable Earth satellite on the Long March 2 was on 5 November 1974, but it was a disaster. The rocket had barely lifted off before it began to sway from side to side, and it had to be destroyed in a fireball by the range safety officer. There was an intensive post mortem which involved analysing telemetry, the wreckage and ground simulations. Due to vibration, the wire from the gyro to the control system had fractured (so it was later determined) and the control system had no basis for stabilising the rocket. It begun to veer out of control, so the range safety officer set off an explosive device in the rocket, destroying the whole assembly a mere 20 s into its first mission – and all because of a single broken wire.

This disaster – China's first – led to a high-level political intervention. Zhang Aiping, who held the lofty title of Minister of the Commission of Science and Technology in the National Defence, descended on the rocket teams to carry out inspections. There was an all-out drive for quality control. All wires were strengthened, or double wires put in. There was a ten-month campaign of more vibration testing of key components to make sure nothing like this would ever happen again. The final assembly of the second rocket began in July 1975, and Zhang Aiping inspected the assembly workshop in person. Not until 20 August 1975 did he give permission for a fresh attempt to be made. Soon after, the next Fanhui Si Weixing cabin left its factory for the long rail journey to Jiuquan. The improvements in the rocket were significant enough for it to be given the designation of Long March 2C. (The 1974 failure became the only launch attempt of the Long March 2 model.)

COASTING TO ORBIT

The second launch attempt was carried out on 26 November 1975, when FSW 0 was launched into orbit from Jiuquan. 7 s after liftoff, the rocket turned toward the south-east, and after 130 s, the first stage engine shut down. The verniers on the second stage ignited, explosive bolts fired to separate the two stages, and the first stage fell to the ground over uninhabited parts of Gansu. The second stage lit up, burning for 112 s. The small verniers continued to fire for a further 64 s as the rocket coasted on towards orbit. The orbital insertion point was 176 km altitude, 1,800 km downrange, over Hunan. Back at the launch site, there was the ever-present memory of the previous year's broken wire. When the report was given that the rocket had entered orbit, several of the engineers sobbed quietly with delight.

As the FSW entered orbit, in Sichuan representatives of the Satellite Tracking and Control Centre arrived from Xian to install radio masts on a number of hills in the region to receive the signals of the cabin, which was due to return three days later – if all went well. Local people were told of the impending arrival from the sky, and the people's militia were mobilised.

The remote ground control station, on the western edge of Chinese territory, sent out the command to the FSW to take up orientation for re-entry and then to fire the retro-rockets. However, the tracking system broke down not long before the moment of retrofire. Ground tracking lost its lock on the spacecraft. Against all safety rules, four brave radio technicians climbed up onto the roof and manipulated the tracking system manually, subjecting themselves to high-frequency radiation in the process. They were not the only people following the satellite carefully; so too was the ageing Chairman Mao, who constantly asked for progress reports.

HELICOPTERS SCRAMBLED

As retrofire approached on the 47th orbit, helicopters were scrambled to watch the cabin come in. The return to Earth was problematical, the cabin being badly burned and the return site far from the intended spot. No one seems to have actually seen it land. Nevertheless, China had succeeded in recovering a capsule at the first attempt, like the Soviet Union many years before. (The United States experienced a dozen failures). No other country has since recovered a satellite from orbit.

The Chinese designated the next series of FSW missions the FSW 1 series, so this series was retrospectively but oddly named the FSW 0 programme, the individual missions being numbered 0-1, 0-2, 0-3 and so on. (Thankfully, it did not have a precursor set, which would, to be logical, have to receive a negative designation).

FSW cabin: final preparations.

First recovery of satellites from Earth orbit

August 1960	United States (Discoverer 13)
August 1960	Soviet Union (Korabl Sputnik 2)
November 1975	China (FSW 0-1)

CRISIS-RIDDEN SECOND MISSION

Following the re-entry problems experienced with FSW 0-1, the cabin was redesigned. The heat shielding material XF was extended to those parts of the shielding which had been badly burnt on the first mission. After several months of intensive work, the new spacecraft was ready by October 1976, timed, according to the Beijing media, to mark the crushing of the Gang of Four.

Launch was set for 7 December 1976, but was delayed when, two minutes before the scheduled lift-off, the swing arm of the rocket gantry failed to retract. Soldiers rushed forward – bravely or recklessly, depending on one's perspective – climbed the tower to the 30 m level, pushed the arm manually back from the fully-fuelled rocket and ran back into the bunker – all in five minutes. Launch director Zheng Songhui approved the launch, which went ahead smoothly. Although there were problems during the mission (the attitude control system nearly exhausted its fuel) it followed the full three-day profile.

Hopeful of a more accurate landing this time, the recovery team mobilised. Four helicopters were scrambled. At headquarters, a plotting map marked the projected descent point while loudspeakers relayed the latest reports. At noon, a sonic boom from the returning cabin rumbled through the valleys of Sichuan. Sharp skywatchers noticed a black dot hurtle in from the north-west, splitting into two. One of the two objects was the discarded heat shield, which was eventually found beside a road, and the other was the cabin. Once the timer activated the parachute, it could be seen gently descending, ending up in a vegetable garden on the side of a hill. One of the four helicopters found a flat spot 100 m away. The crew jumped out, mounted guard, began inspection and removed the previous film.

The third test of the recoverable FSW satellite took place in January 1978, and was also successful. The post-flight announcement confirmed that remote sensing tests had been carried out. The following month, the deputy head of the Space Technology Research Committee, Sun Jiadong, reported on the conclusion of the first phase of the programme. He told his superiors that a new version of the FSW was in design.

IMPROVED VERSION

There was a gap of over four years before the new version appeared. Not only were the spacecraft systems reviewed and improved, but the on-orbit lifetime was extended from three days to five days. A new pointing system enabled the retro-rockets to be fired more accurately; new CCD cameras were mounted to test the possibilities of transmitting data in real time; a radar transponder was added to facilitate recovery; new, more stringent quality control measures were introduced and five air drops were made from 10,000 m, each of them successful. Although new, it kept the FSW 0 designation.

In the improved series, the FSW 0-4 appeared in September 1982. Further missions followed in August 1983, September 1984, October 1985, October 1986 and August 1987 (FSW 0-9). The CCD transmissions were declared to be successful. The October 1985

FSW 0-8 launch, 6 October 1986.

mission took part in a general territorial survey of the land mass of China. By this stage, the missions had become routine. FSW 0-8 was distinguished by coming down in a lake, thus making it the first splashdown in the Chinese space programme, although the lake concerned seems to have been thankfully quite shallow. The 1984–86 missions were land surveys which took more than 3,000 pictures using wide-angle cameras.

INTRODUCTION OF MICROBIOLOGY EXPERIMENTS

FSW 0-9, the last of the series, broke new ground. It was the first mission to fly microgravity experiments and biology tests, and the first to carry a Western commercial payload – two small (15 kg) microgravity experiments for the French company Matra, one of which involved the testing of food growth in orbit. The experimental boxes were handed back to Matra ten days after recovery. A Chinese microgravity experiment was also carried, involving the smelting and recrystallisation of alloys and semiconductors. It is not clear if the final FSW had any remote sensing role at all, or if it was devoted entirely to microgravity experiments.

It is difficult to assess the quality of photographs returned to Earth by the imaging systems of the FSW. Although the Chinese have published photographs of China taken from space, the satellite concerned has never been identified, and in some cases American pictures have been used!

Table 3.2. FSW 0 series, 1975–87

Name	Launch date	Notes
Fanhui Shi Weixing 0-1	26 Nov 1975†	First Chinese satellite recovered from orbit.
Fanhui Shi Weixing 0-2	7 Dec 1976	Second test flight.
Fanhui Shi Weixing 0-3	26 Jan 1978	Third test flight.
Fanhui Shi Weixing 0-4	9 Sep 1982	First operational mission; first 5-day mission.
Fanhui Shi Weixing 0-5	19 Aug 1983	
Fanhui Shi Weixing 0-6	12 Sep 1984	Land survey.
Fanhui Shi Weixing 0-7	21 Oct 1985	Survey of Chinese land mass.
Fanhui Shi Weixing 0-8	6 Oct 1986	Splashed down by accident in a lake.
Fanhui Shi Weixing 0-9	5 Aug 1987	First materials processing, biology mission;; gallium arsenide, monocrystals, plant seeds;; first to fly French commercial payload.

†Launch failure, 5 November 1974

SHI JIAN 2

Shi Jian 1, in March 1971, was China's first scientific satellite, and was highly successful. It was eight years before China was again ready to launch scientific satellites, when an attempt was made to launch three satellites at once. This was by no means unusual, for the Russians had pioneered three-in-one launches in 1964, and had even launched eight-in-one missions (coincidentally, the first taking place the day after Dong Fang Hong was put into orbit in 1970). However, this is not to minimise the Chinese achievement, for the deployment of three scientific packages in this manner can often be accident-prone (as more advanced space nations have sometimes been reminded to their cost).

Project leader was Shi Jinmiao, design director Qian Ji and chief design engineers Wang Zhenyin and Zhu Yilin in the Beijing Institute of Spacecraft Systems Engineering. The original Shi Jian 2 project dated back to April 1972, when it was defined as a space physics satellite to cover eight fields of work. In the course of refinement, three more were added.

Instrumentation for the Shi Jian 2 project

Magnetometer
Semi-conductor proton directional probe
Semi-conductor proton semi-omni-directional probe
Semi-conductor electronic directional probe
Scintillation counter
Long-wave infrared radiometer
Short-wave infrared radiometer
Earth atmosphere ultraviolet radiometer
Solar ultraviolet radiometer
Solar X-ray probe
Thermoelectric ionisation barometer

Shi Jian 2 was a 257-kg, 8-sided, 1.23-m diameter prism, 1.1 m high, with four small solar panels, and its orbit was planned to be 250 × 3,000 km, inclination 70°, with an operational lifetime of six months. It was the first Chinese satellite to store information for later retransmission, and sent back telemetry both in real time and using a tape recorder able to hold 520,000 bits at a time. It had a single ultra-short-wave transmission system which was simpler and lighter than those carried on previous spacecraft, and the information was transmitted to the ground station during daily passes over China. Shi Jian 2 was also the first Chinese satellite to use solar panels (as distinct from solar cells attached to the main body of the spacecraft) and a solar orientation system. As a result, it was able to generate substantially more power than the body-mounted cells of Shi Jian 1. Each of its four solar panels was 1.14 m long and 0.56 m wide – producing a total span of 2.55 m^2 – and each contained 5,188 small solar cells, generating 140 W, which charged nickel–cadmium batteries. The orientation system incorporated a hydrazine-fuelled thruster to rotate the spacecraft at 15–20 revolutions per minute and keep the satellite's panels pointed towards the Sun, thereby obtaining maximum solar power to support the electrical demands of the scientific instruments. Shi Jian 2 also made extensive use of the louvre system of thermal control which was so successful on Shi Jian 1.

NEW CARGO

Shi Jian 2 was to have been the only cargo for this mission, and was almost certainly intended for launch on Long March. However, if the Feng Bao launcher were used there would be sufficient lift capacity to lift two other satellites at the same time. This was quite a complicated exercise, especially since Shi Jian 2 was half complete and work on the other two prospective satellites had not yet even begun. Specifically, it involved the construction of new support structures in the nose cone, small separation rockets and a new range of centre-of-gravity and vibration tests. The change of plan was decided in 1977.

It was essential that the additional two satellites did not unbalance the nose cone or lead to collisions on deployment, or that their radio frequencies interfered with each other. It was a tight fit, for the gap between Shi Jian 2 and Shi Jian 2A was only 4 cm: a clumsy separation would crush the solar panels of Shi Jian 2 and wreck the satellite. The new mission also meant adjustments to the orbit and inclination originally intended for Shi Jian 2. Because of the rush, only two working models of the Shi Jian 2A were made, and one of them was used for the first mission. The Chinese literature, whilst detailed in its account of the Shi Jian 2, has provided virtually no information about Shi Jian 2A and 2B.

FIRST THREE-IN-ONE LAUNCH ATTEMPT

The assembly was brought together for the first time in September 1978. However, the first Chinese three-in-one launch came to grief and failed to reach orbit on 28 July 1979. The second stage vernier engine, designed for the final low-powered thrust to orbit, failed. In September, Ren Xinmin arrived in Shanghai to head up another post mortem, and demanded a complete review of the second stage. In the ensuing changes, the turbo-pump system was rebuilt and more equipment was lifted from Long March 1 to prevent such a setback taking place a third time. The new vernier engine was test-fired six times, on one occasion for 60 minutes non-stop. Even though the three satellites were clearly not to blame in 1979, they were taken apart, and underwent a further three rounds of inspection and testing.

SUCCESS AT LAST

A fresh attempt was therefore organised, and the new rocket was brought to Jiuquan in August 1981. Shi Jian 2, 2A and 2B were put into orbit in darkness at 5.28 a.m. on 20 September 1981, the rocket being lost to sight after 3 minutes. Feng Bao entered orbit after 7 min 20s, separation being achieved in the planned 3.5 s. No fewer than 59 separate operations had to be carried out perfectly in sequence for the separation procedure to work – and it did. The scientific satellites entered similar orbits – 240 × 1,610 km, 59.5°, 103 min. The Shi Jian 2 was actually the back-up model from the failed 1979 launch. (In practice, the battery capacity lost about 6 per cent of its power during the two years spent in storage.) Although they began life in similar orbits, the three satellites were entirely different. Shi Jian 2A was heavier, bell-shaped, with two cones and antennas, and was designed to probe the ionosphere. It is known that Shi Jian 2A explored the ionosphere by transmitting radio signals at 40.5 MHz and 162 MHz to Earth stations. Shi Jian 2B was a combined metal ball and balloon, linked by a thin wire and designed to measure decay rates due to atmospheric drag. The three satellites operated for 332, 382 and six days respectively.

There appear to have been significant scientific results from the Shi Jian satellites. Shi Jian 2 provided details of the configuration, distribution and boundaries of the Earth's radiation belts. By flying during the period of an 11-year peak of solar activity, it was able to measure radiation from our Sun at its most intense.

The mission, incidentally, marked the end of the use of the Feng Bao rocket. Its poor performance, low reliability and the availability of the superior Long March 2C led to its withdrawal from service. The Shanghai workforce was transferred to the ambitious Long March 3, then in the final stages of design, and the Shanghai teams were reintegrated back into the Seventh Ministry.

SHI JIAN 4 AND 5

Only one other satellite received the Shi Jian designation: Shi Jian 4, which was flown on the first flight of the Long March 3A launcher many years later (18 February 1994) (Shi Jian 3, which never flew, was a Landsat (Earth resources) project of the 1980s which never got beyond design stage.) Shi Jian 4 was a 410-kg drum, 1.6 m in diameter, 2.18 m high, with 11,000 2 cm × 2 cm solar cells, whose mission was to study the spatial and spectral distribution of the Earth's charged particle environment. Six scientific instruments comprised a suite of cosmic ray detection instruments: semi-conductor proton and heavy ion detector, static electrical analyser, electrical potentiometer, static single event monitor and dynamic single event monitor. The Long March also carried into orbit an unspecified 1,600-kg payload called the KF-1. Shi Jian 4 entered an orbit of 203 × 261,133 km, 28.6°, period 10.7 hr – one suitable for researching charged particles because of its transit through the Van Allen radiation belts four times a day. It was designed to last for six months, a target which it apparently managed to reach before succumbing to the intense radiation of the belts.

Shi Jian 5, scheduled for the late 1990s, is a project being developed jointly with Brazil. Weighing 398 kg, its purpose is to study the terrestrial magnetosphere. The Chinese have stated that this satellite will become the basis for a range of small, lightweight, low-cost scientific satellites (sometimes called 'smallsats') to be flown over the next few years. This is an American approach in which a basic satellite design or bus is mass-produced, different suites of scientific instruments being attached to the bus according to the type of mission flown.

Table 3.3. Shi Jian series

Name	Date	Remarks
Shi Jian 1	3 Mar 1971	Scientific satellite with cosmic ray detector, X-ray detector and magnetometer.
Shi Jian 2, 2A, 2B	20 Sep 1981†	Three-in-one mission carrying 11 scientific instruments; two other satellites.
(Shi Jian 3)	(Earth observation satellite, not flown)	
Shi Jian 4	18 Feb 1994	Cosmic ray satellite.
Shi Jian 5	Due late 1990s	Magnetospheric satellite.

†Launch failure: 28 July 1979

ASSESSMENT AND CONCLUSIONS

The main achievement of the Chinese space programme in this period was the recoverable satellite series. The Chinese went almost straight from launching a basic satellite to the difficult challenge of orbiting and recovering cabins weighing over 1 tonne. The recoverable FSW satellite involved advanced techniques in space technology, such as heat shields, computers, sophisticated tracking systems and automatic control. The rocket used to support the FSW programme, Long March 2, was a considerable advance over Long March 1, involving new manufacturing techniques, gimballed motors and an inertial guidance system.

In addition to the recoverable programme, the Chinese maintained their commitment to space science. Shi Jian 2, whilst certainly less sophisticated than Soviet or Western scientific satellites of the same period, represented a significant investment in science. Like its predecessor, it appears to have returned a substantial volume of useful information. By contrast, it is difficult to comment usefully on the Ji Shu Shiyan Weixing satellites until the Chinese disclose more of their purpose and function. Even to observers of the classified and obscurer parts of Soviet and American space programmes, the series poses considerable problems of interpretation. It is not clear what purpose is served by retaining the

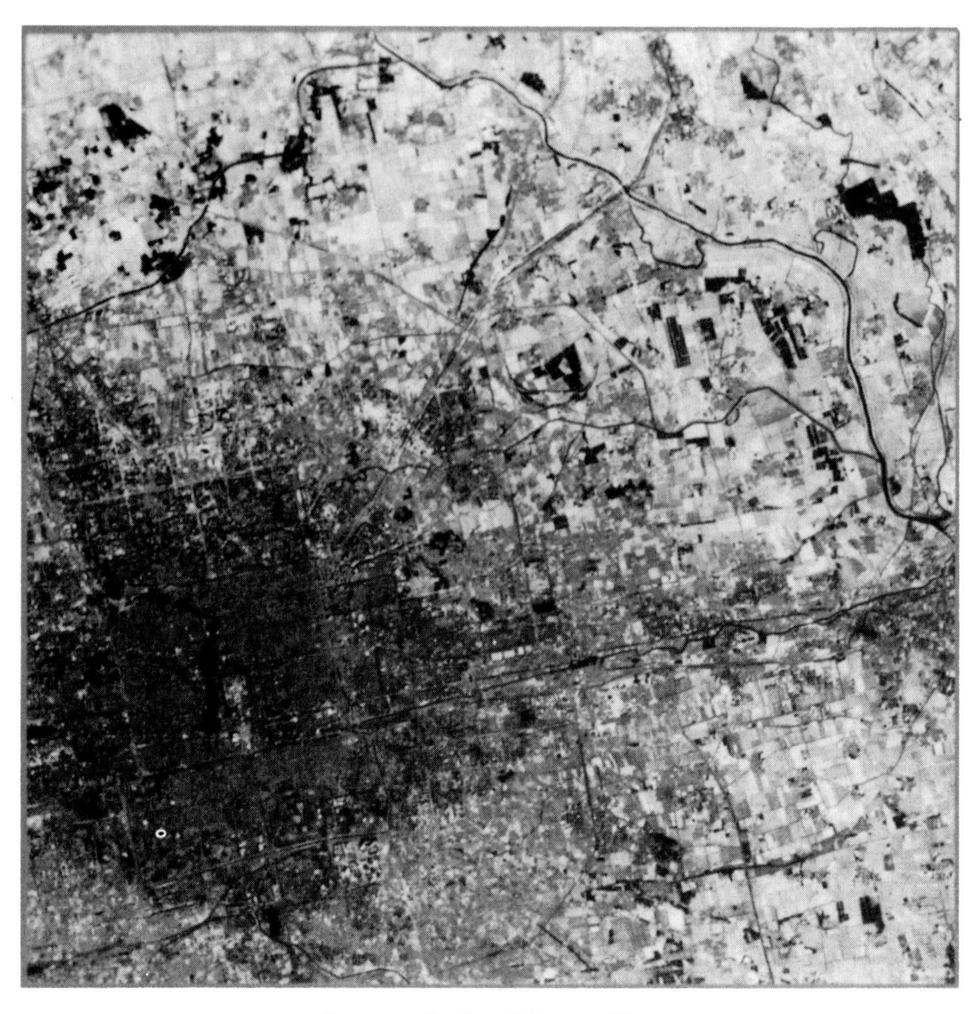

Remote sensing from Chinese satellite.

classification of this programme. As in the 1960s, this decade saw a further period of disruption to the programme because of the unfinished business of the Cultural Revolution. The guiding figure of Zhou Enlai may have been decisive in ensuring the momentum of the programme during difficult times. With the crushing of the Gang of Four in 1976, the four modernisations which followed, and the restoration of political stability in China by Deng Xiaoping, the prospects for more balanced programme development improved significantly.

4

Development of space programmes (1984–1998)

The main achievement of the Chinese space programme in the 1980s was the development of communication satellites. No sooner had China launched its first satellite than it embarked on the challenge of the recoverable satellite. With the FSW recoverable satellite programme well under way, Chinese scientists moved on to a new, ambitious goal: putting satellites into geosynchronous orbit 36,000 km above the Earth. This programme involved the building of a new launcher, the Long March 3, and a new launch site, Xi Chang. Overcoming these difficulties, China then made available its rockets on the world launcher market, with considerable initial success.

The decade of the 1980s and the early 1990s saw the introduction by China of a new series of meteorological satellites – the Feng Yun – and the introduction of new, more advanced versions of the FSW recoverable satellite.

COMMUNICATION SATELLITES: PROJECT 331

The visit of President Richard Nixon in 1972 began the process of international recognition of communist China after years of isolation and siege. Whatever the political significance of his visit, the Chinese were amazed by the satellite television crews who had followed the President's every movement and beamed pictures back live to admiring American homes. This the Americans did through their long-established network of satellites in geostationary orbit.

There were several reasons why China should develop satellite-based communications. Communications satellites offer the possibility of providing advanced telecommunications for a large country quite quickly – quality telecommunication links are now considered an essential element in any modernising country – and they offer both direct television transmission (saving the establishment of elaborate systems of relays) and telephone lines (saving the setting up of land lines) or a combination of the two. In the 1970s, the Chinese leased a number of Western satellite lines to test the potential of a space-based communications system, and they needed no further convincing.

Satellites orbiting at an altitude of 36,000 km above the Earth's equator every 24 hours appear to hover over the same point all the time. The value of such an orbital position was first appreciated by the science writer Arthur C. Clarke, who described its merits in *Radio World* as far back as 1945. The Americans pioneered the use of the 24-hour orbit with Early Bird in 1965, the Russians about ten years later. Now the 24-hour orbit is crowded, and elaborate arrangements exist both for the allocation of slots there, and for ensuring dead satellites are taken out of that orbit and sent to less densely populated regions of the sky (so-called graveyard orbits).

However, the 24-hour orbit presents its own problems. First, to reach an altitude of 36,000 km and enter a circular orbit requires a powerful launcher and upper stage capable of reaching the final destination. Second, the 36,000-km orbit is above the equator, which means that the rocket must not only reach a great altitude but carry out a dogleg manoeuvre southwards. While it is possible to reach 24-hour orbit using a conventional three-stage rocket, placing a more sizeable payload there requires more powerful fuels and/or a restartable upper stage.

The idea of a communications satellite for China was discussed and approved by the Central Committee in 1965, after which exploratory research began. In June 1970, the Chinese Academy of Launcher Technology (CALT) allocated staff for preliminary design studies of launchers and rockets, but a project conference, convened in November 1970, did not have the desired effect in moving the project forward, probably because of the chaos of the Cultural Revolution. Little further progress appears to have been made by 1974[19]. Responding to an appeal from the engineers involved, Zhou Enlai intervened on 19 May 1974 and demanded a report by the State Planning Commission. In the meantime, two further project conferences had been convened: the first by the Chinese Academy of Satellite Technology (CAST) in June 1974, and the second by the Seventh Ministry in September 1974.

MAO'S LAST DECISION

The report of the State Planning Commission (*Report concerning the question of development of this country's satellite communications*) was received by the Central Committee on 25 November 1974, but it seems that it was not given final approval until 17 February the following year. Whichever it was, the decision was not announced publicly until the late 1970s. In April 1975, the Central Committee and then Mao Zedong personally gave the go-ahead to the project (it must have been his last decision on the space programme) and gave it a code name (project 331). Sun Jiadong was appointed chief designer, and Liu Chuanshi director general of the project. Mao's endorsement finally unblocked the logjam. The new rocket was named Long March 3.

The 24-hour orbit presented an immense challenge. The Chinese first considered the development of a low Earth-orbiting system, like the American Telstar, and they also considered the idiosyncratic but highly effective Soviet Molniya system of satellites which orbit the Earth every 12 hours, but with an apogee slowly transiting the northern hemisphere. However, they opted to go straight for a 24-hour system, despite the difficulties.

Because of the complexity of the task involved, in May 1975 the Central Committee and state council set up a project leadership team so as to ensure the proper coordination of all the many different industrial and scientific groups involved. In June, the Chinese

Academy of Space Technology (CAST) held a planning conference which reviewed the progress of global satellite telecommunications to date. Locations for satellites in 24-hour orbit were formally requested from the International Telecommunications Union in March 1977. Locations were sought at 87.5°E, 98°E, 103°E, 110°E and 125°E. The project was listed as a state national priority in September 1977. As was the case with projects 1059 and 651, the government convened a special national conference of scientists, engineers and industry in October 1977 to plan the management and development of the project, the launcher and the satellite. (The conference also took in the launching of the DF-5 and the submarine-launched missile.)

When political turmoil receded, the Chinese invited a high-level American aerospace team to tour Chinese space facilities, discussing with them ways of improving space communications. In 1983, the Italians were invited to establish a 3 m dish ground station in Beijing to receive signals from their satellite, Sirio. For the Chinese, this provided a further learning experience in advance of their own forthcoming comsat project.

The first mock-ups had been built by 1977, the first electrical engineering model followed in 1979, and tests revealed over a hundred problem areas which had to be resolved. Next was a structural model, which was subject to both static and dynamic tests. Temperature tests were run in 1979, the model being cooked in a new hot vacuum chamber called the KM_4 at the Beijing Environmental Engineering Institute. The final, integrated flight version was constructed in the spring of 1983.

Ground testing of Chinese communications satellite.

THE ROCKET

The need for a powerful launcher prompted the Chinese to consider the use of a liquid hydrogen-fuelled rocket. Hydrogen, while having considerable advantages in terms of thrust (50 per cent more than conventional rockets) and environmental friendliness, is a difficult substance to handle. It must be cooled to a temperature of –253° C, and its oxidizer to –183° C. This in turn requires very strong metals, since conventional alloys will turn as brittle as glass under such temperatures. The fuels and oxidizer evaporate quickly on the pad and have to be continuously topped up right to the moment of ignition, and liquid hydrogen has a rate of seepage fifty times higher than water. This area of work is sometimes referred to as cryogenics technology.

The Americans experienced great difficulty with introducing a liquid hydrogen-fuelled upper stage in the 1960s (the Centaur), and the Russians did not bring in such technology until 1987 (Energiya). Not only that, but restarting any rocket stage for a second burn has always proved a persistent problem in rocketry, as the engine must be restarted in zero gravity, without the normal forces which push propellants into the combustion chamber. The Russians had a long series of problems with their Molniya and Proton upper stages failing to restart, and expensive Moon, Venus and Mars probes became stranded in low Earth orbit as a result.

During a feasibility study in 1974, the Chinese weighed up the options of using a conventional launcher and a hydrogen-powered third stage. Whatever the challenges, they decided in August 1976 to go for the most ambitious system – a hydrogen-powered restartable upper stage. However, progress was held up due to the political confusion associated with the rule of the Gang of Four.

CONFIGURING THE LAUNCHER

The Chinese began their first work on liquid hydrogen in the Liquid Fuel Rocket Engine Research Institute in March 1965. The first combustion test was carried out in January 1971, and the first pumps for a liquid hydrogen engine were run in March 1974. Veteran Long March 1 manager Ren Xinmin was appointed Long March 3 project supervisor, and Xie Guangxuan the chief research designer. The new rocket was the biggest yet constructed in China – 43.25 m tall, 3.35 m in diameter and 202 tonnes in weight, with a take-off thrust of 280 tonnes.

The demanding third stage design was assigned to the Chinese Academy of Launcher Technology (CALT), while the first and second stages were given to the Shanghai Institute of Machine and Electrical Design (SIMED). Preliminary designs were approved in March 1978. Responsibility for the control systems of the satellite in 24-hour orbit fell to China's leading radio electronics expert Chen Fangyun (b. 1916). From Huangyan, he graduated in physics at Qinghua University before working in a radio factory in Britain.

NEW ENGINE: INTRODUCING THE YF-73

The new upper stage, named the H-8, required a new engine, the YF-73. This comprised four combustion chambers, with a thrust of 4.5 tonnes, and could gimbal through 25°. The Liquid Fuel Engine Research Institute, which built the YF-73, had conducted experimental work in liquid hydrogen engines since 1965 and had carried out trials in 1971. By 1979, the new YF-73 had passed its stand tests and had proved capable of reignition (at least on Earth). By 1983, the YF-73 had run for 8 hours in 100 tests. These tests were not trouble-free, and the design team experienced problems of over-heating, dangerous hydrogen leaks and materials unable to stand the strain. This was hardly surprising, since the engine turbine turned at a rate of 37,000 revolutions per minute. A particular problem – one which the Americans had encountered with the Saturn V – was the pogo effect, in which the entire rocket was likely to shake up and down with vibration, a snag overcome by the installation of pressure accumulators. In addition to the YF-73, the H-8 upper stage required a set of control thrusters using hydrazine propellant. One of the principal designers of the YF-73 was Wang Zhiren, one of China's few prominent women scientists.

ARRIVING ON STATION: THE APOGEE MOTOR

The transfer to geosynchronous orbit required a complex set of manoeuvres. Several of the key operations would take place when the rocket was well beyond the line of sight of Chinese ground control. Chinese computers lagged far behind Western and Soviet computers in the late 1970s, and China was also embargoed from receiving the latest computer technology. One senior engineer, Hao Yan and a systems specialist, Song Jian, managed somehow to bring the computers up to sufficient standard to function autonomously and carry out the manoeuvres out of tracking range.

A key element in the new rocket was the apogee kick motor – the small, solid-fuel rocket that would be used to adjust the satellite's elliptical geostationary transfer orbit into a circular geostationary orbit. The apogee motor was the third great challenge for the Solid Fuel Engine Research Academy, after the Long March 1 third stage and the FSW retro-rocket. The apogee motor would require a high level of both thrust and precision, as well as being squat in shape to fit into the payload shroud. A total of 45 test runs were carried out between 1978 and 1982, the operational version being delivered to the Long March 3 team the following year.

TROUBLE, LEAKS, EXPLOSION

The three stages of Long March 3 underwent their final all-up tests in summer 1983. The first test in May revealed a serious problem of hydrogen leaks, but round-the-clock re-design of the sealant of the oxygen pump bearings led to a successful test on 26 July 1983. However, it seems that the programme had suffered a serious setback in January 1978 when the motor exploded on the test stand. Buildings were damaged all around, windows were blown out, technicians had their hair singed, and some suffered broken eardrums. This disaster was reported only briefly six years later. Much later still, it was admitted that there had been 'martyrs' – in other words, fatalities.

Designers of the YF-73 liquid hydrogen-fuelled engine

Liu Chuanru
Wang Zhiren
Zhu Senyan
Wang Heng

NEW LAUNCH SITE NEEDED

A related problem to that of the launcher was how to reach equatorial orbit from a launch site with a high latitude (Jiuquan is 41.1° N). Other countries have solved this problem by setting up launch sites on (or near) the equator, as France did in its colony of French Guyana, or, more exotically and very recently, by converting Norwegian oil rigs into launching platforms and towing them to Kiribati in the mid-Pacific (Energiya and Boeing's project Sea Launch). The Chinese landmass lies some distance north of the equator, but by establishing a new site much closer to the equator in southern China, some of the burden of reaching equatorial orbit could be reduced. Accordingly a new site was found at Xi Chang, at 28.25° N (a similar latitude to Cape Canaveral), where the first tests of mockup launchers began in March 1982.

Xi Chang, however, though much closer to the equator, had its drawbacks. The launch site is in hilly country, 1,826 m above sea level, which must have imposed additional construction costs. Climatically, temperatures are more clement than Jiuquan, ranging from −10° C to +33° C. The site has excellent dry weather from October to May, which contributes to an annual average of 320 days of sunshine. However, the weather in the period from May to September is dominated by downpours and thunderstorms. It is also far from deserted, being surrounded by villages. Virtually all the other launch sites in the world are either on the coast (and fire out to sea) or are located in inland desert, thereby reducing to a minimum the risk of civilian casualties. This is not the case with Xi Chang.

THE COMMUNICATIONS SATELLITE

The satellite itself weighed 915 kg, and was a drum 3.1 m tall and 2.1 m in diameter, with the apogee motor underneath and two receiving and broadcasting antennas on top. The satellite was identified subsequently as the Dong Fang Hong 2 series ('2', presumably, in deference to the first Earth satellite, which was '1'). By the time the apogee kick motor had fired, the weight of the satellite on station would be about 460 kg. Its essential function was to receive transmissions from the ground with a high-gain antenna, amplify them and, using two transponders, retransmit them on a spot beam focused on China itself. Solar cells generating 315 W were fitted around the outside of the drum.

The design involved new challenges in devising a satellite which could operate faultlessly for several years, and which could maintain its same unchanging position in the sky and continue to focus its transmitter accurately. The most difficult part was the de-spin system: the satellite itself spun at 50 revolutions per minute (to maintain its stability and ensure that it was evenly exposed to solar rays), but the antenna system had to point in a fixed direction. This involved the development of a reliable, lubricated mechanical de-spin system, one which had caused the Americans much difficulty several years earlier.

The satellites of the Dong Fang Hong 2 series (and those of the subsequent Dong Fang Hong 2A series) carried scientific instruments to measure changes in the intensity of electrons and protons in 24-hour orbit, solar radiation and static electricity on the spacecraft. Typical instruments were a semiconductor and electron detector, semiconductor proton detector, solar X-ray detector and potentiometer.

PREPARING FOR THE FIRST LAUNCH

Zhang Aiping inspected the Xi Chang launch site on 27 October 1983 to check that all would be ready to receive the first Long March 3 rocket and its precious cargo. The new Long March 3 and its payload made their way to the pad on New Year's Day, 1984. The design team had already reached Xi Chang, planning the launch campaign even as their train sped south-westwards. By early 1984, these preparations were complete. The first attempt to launch the Long March 3 was set for 8 p.m. on 26 January, but just after fuelling had begun the guidance platform broke down, and it would not have been possible to control the rocket during ascent. The platform therefore had to be physically removed, and as a result the launch engineers missed their launch window to 24-hour orbit and the rocket had to be drained. This was an early experience of the downside of using liquid oxygen.

Three days later, early on 29 January, the powerful new Long March 3 rocket was once again stacked for its maiden voyage. The storable fuels were loaded 16 hours before take-off, and five hours before lift-off the supercold liquid oxygen and hydrogen were pumped aboard, the amounts being topped up every now and again as wisps of evaporating oxidiser blew away. China then launched its first geosynchronous communications satellite, Shiyan Weixing ('experimental satellite') at 8.24 a.m. It was the first launch of Long March 3 and the first time a rocket had taken off from Xi Chang. All seemed to go perfectly at first, the upper stage, with Shiyan Weixing, entering a parking orbit of 308 × 448 km, inclination 31°. It was intended that the third stage would fire a second time about 50 minutes later to shoot the satellite on its way to geosynchronous orbit, but the engine failed, leaving the satellite stranded in low Earth orbit. Pressure in the launching chamber reached only 90 per cent of the level necessary and after 3 seconds it collapsed. This was a great disappointment, and it must have been little consolation to the Chinese that this was one of the most frustrating problems in astronautics, which had equally taxed the other space powers.

The Chinese decided they would salvage what they could of the Shiyan Weixing mission. They separated the satellite from the third stage and conducted the final on-orbit manoeuvres that should have been carried out at 36,000 km, and in the event the apogee motor moved the satellite into an orbit of 400 × 6,480 km. They also made the satellite carry out the range of station-keeping adjustment manoeuvres that would have been necessary had the comsat entered the intended orbit. The satellite's various communications systems were tested rigorously, being turned on and off and tried in different régimes. The satellite experienced a number of problems in its electrical supply and heating systems, which were remedied in good time for the next model.

BEATING THE SUMMER THUNDER

A second Long March 3 was already at Xi Chang, and ground tests were made to try to resolve the problem. There was added urgency to do so, since summer storms affected Xi Chang from May onwards. The launch teams worked day and night, right through the spring festival, and fax lines between Xi Chang and Beijing ran hot as launch teams and design teams tried to diagnose the fault, working against time. To test for problems, the third stage was hot fired again at the Liquid Fuel Rocket Engine Testing Station in Beijing, the tests being completed on 20 March. The modifications had to be performed on the new rocket, which was already stacked on the launch pad. Wearing breathing apparatus, a team of engineers, led by Ni Zhongliang and Ciu Yajie, had to crawl into the third stage from the gantry. There was room neither to sit nor to stand and they had to work at awkward angles in the plumbing system, always being careful not to damage any delicate machinery. The work took them three days.

The next mission was flown four months later, on 8 April 1984. The new *Yuan Wang* comship took up position, 1,000 km off the Chinese coast. The satellite was called Shiyan Tongbu Tongxin Weixing ('experimental geostationary communications satellite'). Despite their best efforts, the first of the spring thunderstorms had already arrived in Xi Chang, a potential cause of disaster should the launch of a hydrogen-fuelled rocket be attempted. In the afternoon, Xi Chang launch site clouded over and it began to rain. The launch controllers went into emergency conference. The forecast predicted that the clouds would clear for the intended launch time, 7 p.m. An elderly local village sage stepped forward and predicted clear weather for the intended launch time 'or I'll never take a drink in my life again!' he swore. Now certain of fine weather, the controllers ordered the countdown to continue.

At dusk the clouds duly rolled away, and twinkling stars could just be discerned. Spotlights played on the waiting rocket. Fuelling began. About 50 minutes before launch, crowds began to gather on the surrounding hills. By 7 p.m. the last fuels had been loaded and the trucks dispersed. The engines were armed. At one minute before launch, the hilltop spectators could see the attachment plugs fall from the side of Long March. The electric cable arm retracted. At 19.20.02 a technician pressed the red firing button and Long March 3 headed skywards. But had the engineering team carried out the job?

As Long March 3 upper stage cruised southbound over the Earth's equator, the rocket reignited and hurtled the 1-tonne satellite onwards. Twenty minutes later, tracking ships in the Pacific reported that this time the burn to geosynchronous orbit appeared to be correct. The satellite's own motor fired to settle it into geosynchronous orbit on 12 April, and trimmed the orbit four days later. But would it work? These were anxious moments. As the satellite drifted into position, the thermal control in the battery system broke down and the current began to fluctuate alarmingly. In a risky repair, the ground controllers in effect told the spacecraft's computer that the satellite had completely broken down and that it must make a full recovery, rather like restarting a computer system by turning it off and beginning again. It worked. At 18:27:57 on 16 April the satellite arrived on station at 125°E longitude. The spacecraft was spun, the de-spin system turned on and transmissions begun. The first relay tests were carried out the next day, and an hour's television was transmitted to the most distant regions of the country. The first pictures were clear, and stable, the colours realistic and the sound well up to standard.

ZHANG AIPING'S TELEPHONE CALL

Zhang Aiping was one of the first people to test out the new comsat by making a much-publicised telephone call to a distant party committee in Xinjiang. The satellite established China's first 200 satellite-based telephone lines, connecting Beijing with Urumqi, Lhasa, Hohhot (inner Mongolia), Chengdu (Sichuan) and Guangzhou. The voice quality was good, with almost no background noise or interference. The satellite was formally handed over to its new telecommunications owners on 25 April for tests, and the system was declared operational on 24 May.

To mark the occasion, on 30 April a solemn conference, presided over by Hu Yaobang, was convened in the Great Hall of the People in Beijing. Subsequent accounts suggest it was less than solemn; the Chinese had much to celebrate, as the achievement had placed their country in the top league of applied space engineering. Ten years of hard work had been vindicated, though the official history noted that several of the engineers had gone grey, some had suffered ill-health from overwork, and others had even passed on prematurely[20].

NEW LANGUAGES FROM SPACE

All went well with Shiyan Tongbu Tongxin Weixing. Small thrusters were occasionally used to readjust its orbit, and although designed for only a short working life it worked perfectly for more than four years until it was taken out of service. Its 15 radio and television channels transmitted programmes in Cantonese, Amoy, Hakka, Japanese, Spanish, Russian, Burmese and Tagalog, some of these languages not hitherto familiar on the international satellite network.

Following the success of Shiyan Tongbu Tongxin Weixing, the Chinese proceeded to the launch of the first operational geosynchronous communications satellite, Shiyong Tongbu Tongxin Weixing ('*operational* geostationary communications satellite'). The main difference from its predecessor was an improved 0.7-m diameter dish to beam its transmissions to Earth. It was launched two years later on 1 February 1986, reaching geosynchronous orbit two days later and its final operating position at 103° E longitude on 18 February. Its motor was fired monthly to ensure it stayed precisely in this position, and it operated for four years, transmitting to 30 television stations, until it drifted off station. It was the last in the Dong Fang Hong 2 series.

DONG FANG HONG 2A: 3,000 TELEPHONE CALLS AT A TIME

Following this, Long March 3 was used for a series of commercial launches for other countries (these are described in the next section). The next 24-hour domestic satellite saw the introduction of the Dong Fang Hong 2A series. The Dong Fang Hong 2A comsat was 3.68 m tall, weighed 441 kg on station and had a design life of four years, power being supplied by 20,000 solar cells. It had four transponders able to transmit five television channels and 3,000 telephone calls at a time. The first, Shiyong Tongbu Tongxin Weixing 2, was launched on 7 March 1988, took up position at 87.5° E and doubled its design life by being still operational eight years later. Equally successful were Shiyong Tongbu Tongxin Weixing 3 that December (110.5°E) and Shiyong Tongbu Tongxin Weixing 4 two years later (98.5°E). Its launch, on 4 February 1990, was observed by Chinese premier Li Peng. These early satellites were used to achieve complete television coverage for China, with 30 channels, and to permit telephone and fax services to be sent by satellite for the main governmental agencies and development bodies. Around 30,000 receiving dishes were built, with education programming going out for many hours a day, and reaching over 30 million people.

The run of successes came to an end on 28 December 1991, when a Long March 3 launched what should have been Shiyong Tongbu Tongxin Weixing 5. Rather like the first attempt to send a satellite to geosynchronous orbit in 1984, the third stage failed, this time after burning for 58 s. Apparently, the helium pressurising gas in the third stage sprang a leak, and pressure in the combustion chamber fell to zero 135 s into the burn. Once again, the Chinese separated the payload in an attempt to salvage something from the mission, though it is not known if they achieved anything practical from the manoeuvre. In the event, the following year the Chinese space authorities bought an American comsat, Spacenet 1, which was already in orbit and nearing the end of its useful life. Motor firings were used to move the Spacenet from its location at 240° E to what may have been the intended destination of Shiyong Tongbu Tongxin Weixing 5 at 115° E. They then renamed it Zhongxing 5 (Zhongxing means 'the star of China'), though Zhongxing 1–4 were never retrospectively identified (although they could have been Shiyong Tongbu Tongxin Weixing 1–4).

DONG FANG HONG 3 SERIES: 8,000 TELEPHONE CALLS AT A TIME

The third generation of Dong Fang Hong communications satellites was already in the pipeline. The purpose of the new generation was to increase twelvefold the capacity of the previous series and guarantee a working life of eight years. It was also China's intention to use a design called the DJS platform, which could be adapted for other satellites. The Dong Fang Hong 3 series, unlike its predecessors, had a minor (20 per cent) Western design contribution. In July 1987, agreement had been reached between the Chinese company responsible, the Great Wall Industries Corporation, and Messerschmitt Bolkow Blohm (MBB) of Germany for the development of the Dong Fang Hong 3. MBB was responsible for part of the design, besides the solar array and the antennas.

Dong Fang Hong 3 has double the weight of its predecessors – 1,145 kg on station – is 5.71 m tall and has a 2-m diameter communications dish with six spot beams. It has 24 transponders which can transmit six colour TV channels and take 8,000 telephone calls at a time – able to cover 90 per cent of China – and has a working life of eight years. The solar wings have a span of 18.1 m and are able to generate 2,000 W. The assembly engineering team was led by Li Chunqi, a 48-year old engineer who passed up a university place for a chance to work on satellites. The welding of the Dong Fang Hong fuel tank involved the use of considerable electrical power supplies, to the extent that electricity was turned off in the small town in north-west China where it was being built in order to guarantee electrical power for the welders. The citizens were duly warned of the power cuts, and the engineers, expecting a less than enthusiastic response, especially in mid-winter, anticipated the worst from irate citizens the following morning. Instead, to their surprise, callers to the factory gates were only interested to know if the welding job had been accomplished (which it had).

However, all did not go well on the first mission. On 29 November 1994, the new Long March 3A rocket left it in its transfer orbit of 181 × 36,026 km. The Chinese used its propellant over time to raise the perigee to 35,181 km by 29 December. By the time it reached that altitude, all the propellant had been used up and the satellite had to be abandoned. The problem is understood to have been in the satellite, not the launcher.

Dong Fang Hong 3 comsat.

The Chinese, then engaged in renaming their communications satellites, called it Zhongxing 6.

After the failure of Zhongxing 6, the Chinese bought a Hughes 276 spacecraft from the United States. They launched it on one of their own Long March 3 rockets, but once again the transfer manoeuvre to geosynchronous orbit went wrong and it became stranded between 100 km and 17,230 km. Apparently, the pressurising gas failed, which caused the thrust to stall a mere 48 seconds before the satellite would have reached orbit. Zhongxing 7 was then abandoned. It was the third successive failure in five years. The Chinese Telecommunications and Broadcasting Satellite Corporation, which was to have operated Zhongxing 7, was eventually paid $25.9 million by an insurance company for the loss.

Eventually, however, a Dong Fang Hong 3 satellite reached orbit successfully – put up on a Long March 3A on 8 May 1997. The satellite, designed to replace Zhongxing 7, was built by the Chinese Academy of Space Technology in Beijing, with assistance from Germany's Daimler Benz Aerospace. Just to confuse an already muddled identification system, the Chinese dropped the Zhongxing system and called it the Dong Fang Hong 3-2. On-orbit testing of its systems was successfully completed at the end of August 1997. Eventually, it too will be phased out in favour of the 3.5-tonne Dong Fang Hong 4 series in 2000. The Dong Fang Hong 4, using the DJS-2 common platform, will have 24 transponders, support high-speed data links, generate 6,000 W of electrical power and operate for up to 15 years.

Television ground receiving dish.

BENEFITS OF COMSATS

By 1996, China claimed that the beams from its satellites were reaching 83 per cent of Chinese people. Some 1.37 million college students had been trained on distance learning programmes beamed down by China's comsats. China's development of its communications networks gathered pace during the 1990s; speed has been more important than political imperatives. In addition to its own networks, China has pragmatically bought communications satellites from the West and often leased lines on Western satellites, rather than wait to establish a completely indigenous service. In 1991 for example, the Spar communications group won a $36 million contract to provide ten new Earth terminals using state-of-the-art 13 m diameter dishes operating on Intelsat's Indian Ocean network, providing the highest quality communications in China's new economic development zones.

By 1996, China routinely routed 8,000 domestic and 25,000 international calls via satellite. The Bank of China transferred data and had dealings with 350 branches by satellite. There were even proposals to use satellites to control and monitor rail traffic in such a way as to double the usage of the network. China estimates that it will need about 150

Table 4.1. First domestic satellites to geosstationary orbit

Name and type	Launch date	Notes	Location
Shiyan Weixing (Dong Fang Hong 2-1)	29 Jan 1984		Stranded
Shiyan Tongbu Tongxing Weixing (Dong Fang Hong 2-2)	8 Apr 1984	Experimental 24-hour comsat.	125° E
Shoyan Tongbu Tongxing Weixing 1 (Dong Fang Hong 2-3)	1 Feb 1986	First operational 24-hour comsat.	103° E
Shiyong Tongbu Tongxing Weixing 2 (Dong Fang Hong 2A-1)	7 Mar 1988	First of Dong Fang Hong 2A series.	87.5° E
Shiyong Tongbu Tongxing Weixing 3 (Dong Fang Hong 2A-2)	22 Dec 1988		110.5 E
Shiyong Tongbu Tongxing Weixing 4 (Dong Fang Hong 2A-3)	4 Feb 1990		98.5° E
Shiyong Tongbu Tongxing Weixing 5 (Dong Fang Hong 2A-4)	28 Dec 1991	Third stage failed to fire; stranded in low Earth orbit.	Stranded
Zhongxing 6, or Dong Fang Hong 3-1	29 Nov 1994	Inserted in wrong orbit; eventually reached station but abandoned.	Intended for 125° E
Zhongxing 7	18 Aug 1996	Bought from US; stranded in wrong orbit; abandoned.	Stranded
Dong Fang Hong 3-2	8 May 1997		125° E

transponders in orbit on comsats by 2000 (each satellite has between 24 and 36 transponders) in order to meet the high level of domestic demand for telephone, television, data links and mobile communications. At present, 80 per cent of China's space communications goes through foreign comsats, and only 20 per cent via their own, a proportion the Chinese would like to reverse.

COMMERCIALISATION OF THE CHINESE SPACE PROGRAMME

Following the successful launch of the Long March 3 in 1984, China began to offer its Long March series of launchers to the West. The decision to do so was strongly driven by the reduction in space spending after 1978, which was accompanied by an authorisation for the space agencies to generate as much external income as they could from abroad. The formal announcement was made by astronautics minister Li Xue in October 1985. The Chinese intention was to interest Western communications companies in using the Long March to get their comsats into geostationary orbit. The offer generated little media interest at the time. Western companies did take up more modest opportunities to fly dedicated payloads on recoverable satellites, and in the course of 1987–89, China flew commercial microgravity experiments for France (FSW 0-9) and Germany (FSW 1-2). Further opportunities were publicly advertised in the western press[21], and the Chinese issued a *Long March 3 user manual* to encourage familiarity with their product. In 1985, at Zukiha, near Tokyo, Japan, the Chinese contributed their first stand to an international science and technology exhibition, showing off the new Long March 3 launcher. Despite the lack of an immediate response, China continued to press its offer.

Little happened to the launch offer until 1986, when, in the space of a few months, America lost the space shuttle *Challenger* and two of its other leading rocket launchers (a Titan and a Delta) spectacularly exploded. To compound the problem, Europe's Ariane also went down. Western companies in a hurry to launch satellites faced increasing delays and were forced to turn to Russia or to China, whose capabilities they knew little about.

RESTRICTIONS, PRICES AND QUOTAS

The Chinese, however, may not have reckoned with the battery of trade and defence sanctions at the command of the world's economic superpower. Both the Chinese and the Russian commercial space programmes were subject to American military restrictions, ostensibly designed to prevent them copying Western satellites *en route* to the launch pad. These restrictions were enforced through export licences and what was termed the CO-COM agreement between the United States, the North Atlantic Treaty Organisation and neutral Western countries. The Americans required any American company to obtain permission to launch a satellite on a Chinese launcher. The same requirements were equally enforced on non-American companies contemplating the use of American communications satellites. The Chinese tried to meet American concerns by guaranteeing an inspection-free transit of the satellite from the United States to the top of the launcher.

Even when American-built equipment was cleared, from the military and security point of view, the Russians and Chinese faced trade quotas on how many satellites they could actually launch each year and the price to be offered. The prices proposed by the Chinese for Long March 3 launches to geostationary orbit have generally been below that for Ariane (Europe) and the American launchers, but above the Russian (Proton). Under a Sino–American agreement signed in January 1989, for the period 1988–94 the Chinese were permitted nine commercial satellite launches. China had to agree to United States demands to 'ensure that the market was open and fair'[22]. This quota was extended to 11 launches for 1995–2001, provided that China did not offer prices less than 15 per cent below Western rates[23]. Whilst lecturing the communist and former communist nations about the virtues of free enterprise, the United States explained the need for these quotas as being to permit 'disciplined Chinese participation in the market'. At the time, China had about 4 per cent of the world market.

Despite the agreements, there were repeated allegations in the international trading world that China's launch prices were undercutting Europe and the United States. Such altercations were and are by no means unusual in international commerce, and they affect many other areas of trade (for example, aircraft). The United States was equally critical of the pricing policy for the Ariane launcher, though the difference was that the Europeans were better able to stand up to American economic pressure. Pricing in the international launcher business is in any case far from transparent, since all the main launch companies have had their development costs underwritten by their governments and are effectively guaranteed the launches of many of the satellites manufactured in their respective regions, especially the military ones. To complicate matters still further, several companies vying for the launcher business often offered the first flight for a customer at an 'introductory rate'. The company offering Ariane also routinely provided a range of financial packages. These rates became the subject of much contention. The Chinese have been very much aware that they undercut Western prices, but they argue that they have lower labour costs and lower prices for raw materials, and that government investment for the programme is now very low, forcing space companies to generate their budgets abroad.

EXPORT LICENCE CRUX

For an American-made satellite to fly on a Chinese rocket, an export licence was required. Because most of the world's comsats are made in the United States, this necessarily involved foreign countries like Australia pleading to the United States for permission to fly American satellites on Chinese launchers. Normally, export licences are decided by the State Department alone, but in these cases reference was also made to the National Security Council and the Economic Policy Council. When deadlock ensued, the matter was referred upwards to the White House. Secretary of Defence Casper Weinberger (who visited Xi Chang) and his successor Frank Carlucci both took the view that the risks of technology transfer to China were minimal and that the whole matter essentially concerned political and economic considerations. In the former case, political considerations revolved around the desirability of doing business with a political system which many Americans repudiated. On the economic front, considerations revolved around arguments for free trade (favoured by the satellite manufacturers) and the protection of the domestic American launcher market (favoured by American launcher companies). Thus the question of export licences for the Chinese became a domestic lobbying battleground for the giants of American industry.

Long March 2E lifting off by night.

The outcome of this domestic American political battle was that President Reagan ruled in favour of granting export licences[24]. These licences were renewed by his successor, President Bush, though they were temporarily withdrawn in 1989 (due to political difficulties following the summer events in Tiananmen Square) and again in 1991 (to put pressure on China to sign international agreements on the proliferation of missiles).

The Chinese offered several variants of their Long March – the 2C, 2E, 3, 3A, 3B and 3C – and put forward a range of options between leaving the satellite in low Earth orbit, low Earth orbit for a subsequent 24-hour orbit (with the Western company providing its own rocket to ensure transfer to geostationary orbit), right up to deployment in geostationary orbit. To complete the package, Chinese insurance companies also offered insurance to the foreign customers.

FIRST COMMERCIAL MISSION

The first commercial mission, Asiasat 1, in April 1990, was entirely successful. Asiasat 1 was owned by a Hong Kong company, one which included Chinese (Citic Technology Corporation), British (Cable and Wireless PLC) and local financial interests (Hutchison Telecommunications), formed in 1988. Asiasat 1 had a history. It had originally been launched into orbit as Westar 6 by the space shuttle in February 1984, but had been stranded in low Earth orbit when its motor failed to fire. Nine months later, American shuttle astronauts on mission 51A had retrieved the errant Westar and returned it to Earth for relaunch. The much-travelled but still unused Westar had then been bought by the Asiasat company, and this was its second – and more successful – journey into space. They paid China $30 million.

The second commercial launch was likewise a success – a Pakistani satellite called Badr, riding the Long March 2E on its maiden voyage. Badr was a 52-kg experimental satellite built by the Pakistan Space and Upper Atmosphere Commission, and was the first Pakistani satellite to enter orbit. China was paid $300,000 for the launch.

A Swedish satellite, Freja, rode into orbit in October 1992, accompanying the FSW 1-4 recoverable satellite mission. Named after the Viking goddess of fertility, Freja was a 259-kg spacecraft commissioned by the Swedish Board for Space Activities. A small American upper stage was used to place it in a high orbit, from where its seven experiments could study the northern hemisphere's aurorae, electrical and magnetic fields, particles, plasma, electrons and magnetosphere. The spacecraft also carried an experimental communications payload called Mailstar. Sweden paid China $4.3 million for the lift.

FIRE ON THE PAD!

The trouble started on 22 March 1992, with near disaster on the pad when the launch of Optus B-1 was aborted. (The Australians paid $15 million each for two Optus launches – in effect a promotional rate – and the launch of Optus B-1 was commissioned by an Australian communications company, Aussat.) The Long March counted down to lift-off and the main engines ignited, but after 3 s they were turned off by the computers, which detected a fault a mere fraction of a second before the scheduled lift-off. However, three of the four restraints holding the rocket to the pad had been released by this stage, and fires

Long March 2C lifts off carrying the Swedish satellite Freja.

had already broken out at the base of the rocket. Reckless for their own safety, ground crews rushed forward to douse the flames and turn off the ignition systems. In their hurry, none donned oxygen masks and all inhaled the toxic gases swirling around the launch pad. The entire team had to be hospitalised, but all survived, though many had vomited blood from poisoning. The rocket was not safed until 39 hours later, when it was drained of all its propellants.

Optus B-1 was, however, eventually launched properly several months later on 13 August. A hundred Australians and other visitors attended the launch in Xi Chang, and the event was broadcast live on Chinese central television. Cameras switched between the rocket lifting into the early morning sky and its hopeful customers eagerly watching it climb skywards. The purpose of the Aussat system was to provide radio and television services for remote areas of Australia, air traffic control and educational and medical services by television.

Optus B-1 takes to the skies, 13 August 1992.

DEBRIS, RECRIMINATION

The earlier difficulty was nothing compared to what happened with Optus B-2 on 21 December 1992. Optus was launched with an American kick motor, the Star 63F, to place it into geostationary orbit. Some 70 s into the mission, a cloud of gas could be seen emerging from the shroud at the top of the launch vehicle. The remainder of the launch proceeded normally, but there was widespread consternation when it transpired that all that had reached orbit was satellite wreckage. It seems that the satellite met with an accident about a minute into the mission, but that the shroud had contained the explosion.

There was considerable recrimination afterwards between the Chinese, the Australians and the United States as to who was responsible. The Americans blamed the Chinese for a faulty shroud which opened under pressure; the Chinese blamed the Americans for failing to attach the satellite to its upper stage sufficiently to withstand vibration. Western press coverage laid the blame firmly on the Chinese. In the end, both the Chinese and the Americans agreed to paper over the cracks, eventually issuing a joint statement to the effect that neither the launcher nor the satellite was to blame!

The Chinese recovered somewhat with two successful launches in 1994: Apstar 1, flown for a Hong Kong company in July, and Optus B-3 in August, effectively replacing the satellite which had been destroyed two years earlier.

LONG MARCH CRASHES IN FLAMES

The new confidence did not last long, however. On 25 January 1995, another Hong Kong satellite, Apstar 2, also carrying a Star 63F kick motor, was lost. At 51 s into the mission there was a catastrophic explosion, and the entire launcher and satellite were lost. Television pictures showed the rocket crashing in a ballooning cloud of red, yellow and black toxic smoke. Six villagers died and 23 were injured. The mission was an insurance loss of $160 million, the premium being 18.5 per cent. Mysteriously, the explosion appeared to start at the top of the rocket, not the bottom part which was actually firing at the time. The Long March was grounded while the problems were rectified, and more recriminations flew back and forth. In a repeat performance of what had happened the previous year, the next two launches went smoothly: Asiasat 2 (Hong Kong, November) and EchoStar (United States, December). Asiasat used a small Chinese final stage, the EPKM kick motor, to reach final orbit.

The causes of the two satellite losses – Optus B-2 and Apstar 2 – were never satisfactorily resolved, either by the launching company or by the customer. Neither the American nor the Chinese accident investigation reports were published. The most plausible explanations for the two failures have been published by Clark, who has put forward two possibilities[25]. One is that the manner in which the American comsat should be placed on the top of the Long March has been miscalculated, and that it is unable to stand the high air pressure which rockets experience about a minute into flight. Between 45 s, when a rocket goes supersonic, and about 90 s, it experiences maximum dynamic pressure – the time when it is put under the greatest structural strain (after that, the air is thinner and the pressure diminishes). The alternative explanation is that there is a problem with the American upper stage kick motor, the Star 63F, which has on both occasions exploded about 70 s into the mission. Whichever be the case, the problem is something for both countries to remedy, but blaming the Chinese launcher alone is probably the least helpful approach.

SAINT VALENTINE'S DAY MASSACRE

Then disaster intervened once more. 14 February 1996 saw the launch of a $60 million American Intelsat 708 advanced communications satellite. It was the first flight of the

Long March 3B, a new version of the Long March 3 able to lift a record 5 tonnes to geostationary orbit. Although a new version, it relied heavily on well-tested rockets: the main stages were essentially those of the Long March 3A while the strap-on rockets had been verified on the Long March 2E. However, because the Chinese had received less than they had hoped for in launch fees, they did not have the resources to make a test flight of the 3B before committing the new rocket to its first commercial mission. This decision proved calamitous.

Ground controllers were horrified as, a mere 2 s after liftoff, the rocket began to tilt to one side, turned sideways and exploded with an enormous bang 22 s later, 1,500 m away, showering debris for miles around. There were two fatalities, and a further 80 people were injured (although later figures quoted as many as 56 fatalities). The crash was so shattering that no large pieces of debris were ever found. It was a very visible failure, screened instantly throughout the Western world and provoking much comment about the temperamentality of Chinese rockets. Whatever might be said about the Optus B-2 and Apstar 2 failures, this time no one could argue but that there was a fault in the launch vehicle.

Western investors and insurers later called the episode the Saint Valentine's Day massacre. Two investigating committees were appointed and international experts invited to join. Four possible causes of the failure were indicated, including a broken wire supplying electrical power to the guidance system. By the end of the month, the China Great Wall Industry Corporation stated that the guidance platform had gone badly wrong, causing the accident. Another computer guidance problem was to cause the equally spectacular loss of Europe's brand new Ariane 5 on its maiden voyage less than four months later.

LOSS OF CONFIDENCE

With the crash of the Long March 3B, Western investors lost confidence in the Chinese launcher system. At one stage, China had been heading for a 9 per cent share of the international launch market. Now, satellites due for launch on Chinese rockets became uninsurable. Three American companies at once transferred their payloads to Atlas and Ariane launchers, as did an Argentinean company. Some customers went to the Russian Proton, which, after many years languishing in the commercial doldrums, was now enjoying a long order book.

China continued its efforts despite these difficulties. Five months after the Intelsat disaster, in July 1996, the Long March 3 put Apstar 1A into its proper orbit. Then problems arose again. This time, only the following month, a Long March 3 was carrying an American-built communications satellite for the China Telecommunications Broadcast Satellite Corporation, the domestic communications supplier. During the burn to geostationary orbit, the third stage lost pressure and shut down 48 s early, leaving the Chinasat 7 satellite stranded only half-way to its intended destination. The satellite was insured domestically for $25.9 million, which was paid up by the China Assets Insurance Co. that October.

Following this further failure, the Chinese instituted a rigorous programme for greater quality control and launch safety[26]. The programme was put under international quality standards (the ISO 9000 quality mark), there were additional quality checks, the guidance system was redesigned, and arrangements were made to evacuate areas near the pad as lift-off neared. A quality-control company, the New Decade Institute, was called in, and a 28-point quality control system was adopted, as was a 72-point regulation for management and production. The Chinese insisted that an international team of French, German and British experts approve the reforms, which obviously paid off; on the early morning of 20 August 1997, the Long March 3B eventually made its debut, lofting a comsat for the Philippines.

Table 4.2. Chinese commercial launches

Name	Customer/ country	Launch date	Notes
Asiasat 1	Hong Kong	7 Apr 1990	First commercial launch; for Hong Kong company.
Badr	Pakistan	16 Jul 1990	Low Earth orbit.
Optus B-1	Australia	13 Aug 1992	First for exclusively Western company; followed abortive launch attempt in March.
Freja	Sweden	6 Oct 1992	Flown on FSW 1-4 mission.
Optus B-2	Australia	21 Dec 1992	Explosion in upper part of rocket.
Apstar 1	Hong Kong	21 July 1994	Successful.
Optus B-3	Australia	13 Aug 1994	Successful.
Apstar 2	Hong Kong	25 Jan 1995	Exploded 51 s into flight.
Asiasat 2	Hong Kong	28 Nov 1995	Successful.
Echostar 1	USA	28 Dec 1995	Successful.
Intelsat 708	USA	14 Feb 1996	Crashed immediately after launch.
Apstar 1A	Hong Kong	3 Jul 1996	Successful.
Chinasat 7	China (domestic communications)	18 Aug 1996	Stranded.
Agila 2	Philippines	20 Aug 1997	First successful flight of CZ-3B.
Apstar 2R	Hong Kong	16 Oct 1997	Successful.

FUTURE COMMERCIAL FLIGHTS MANIFESTED

Some customers, however, did stay, including the Motorola Corporation, which had booked a series of Long March 2C launches to low Earth orbit for 22 satellites of its revolutionary new global communications system, Iridium. (Great Wall has a 5 per cent interest in the Iridium project.) The Long March 2C is one of China's most reliable launchers, with a success rate of 100 per cent, and was adapted with a special dispenser (SD) for the Iridium system, being named the CZ-2C-SD. The Motorola Corporation contracted China to launch 22 satellites in the 66-satellite Iridium constellation of comsats (the others being selected to fly the American Delta and the Russian Proton). A test of the SD system was made on 1 September 1997 when the Long March 2C-SD, flying out of Taiyuan launch site discharged two mock-up Iridium satellites at appropriate points in its orbit, clearing the way for the first operational mission. The first real mission came on 8 December 1997.

Despite the launch disasters of the mid-1990s and the decision of several companies to move their payloads to other launchers, China continues to have a manifest for commercial satellite launches, though this does not necessarily mean that all are yet insurable. Most involve the Long March 3B, which had the disastrous maiden flight of February 1996.

Table 4.3. Commercial flights manifested

Name	Company	Launcher
Sinosat 1	China Aerospace Corp. and Euraspace for People's Bank of China	CZ-3B
Chinastar 1/ Zhongwei 1	China Eastern Communications Satellite Ltd with Lockheed Martin	CZ-3B
Iridium (× 4)	Motorola	CZ-2C-SD
Chinasat 8	Loral Space and Communications Co., USA	CZ-3B

In June 1997, the Great Wall Industry Corporation signed a protocol with the Hughes Corporation of the United States for China to launch 10 Hughes comsats over the period 1998–2006. This deal followed two successful Long March launches in May and June (Dong Fang Hong 3-2 and Feng Yun 2). This substantial order must have done much to restore Chinese confidence in its commercial space programme. In August 1997, nine Chinese property insurers came together to form an insurance syndicate for future comsat launches.

The prices paid by commercial customers for the Long March rocket depend, of course, on the requirements of individual customers and missions. Table 4.4. shows estimated prices expected for Long March launches. In comparison, costs for geostationary transfer orbit for the European Ariane 4 launches range from $65 million to $110 million, depending on the requirement and payload. Russian Proton launch costs are in the $60 million range, and the American Delta 2 between $45 million and $50 million.

Table 4.4. Long March capabilities and prices

Launcher	Capability	Price
Long March 1D	1,000 kg to low Earth orbit.	$12 m
Long March 2D	3,300 kg to low Earth orbit.	$20 m
Long March 2E	2,494 kg to geostationary transfer orbit.	$50 m
Long March 3	1,400 kg to geostationary transfer orbit.	$40 m
Long March 3A	2,500 kg to geostationary transfer orbit.	$45 m
Long March 3B	4,800 kg to geostationary transfer orbit.	$70 m
Long March 4	4,000 kg to low Earth orbit; 1,500 kg to polar, Sun-synchronous orbit.	$30 m

METEOROLOGICAL SATELLITES (FENG YUN)

The development of weather satellites was logical for the Chinese. Accurate weather forecasting has always been important for a large country so dependant on agriculture in an area vulnerable to damaging storms and floods. The United States launched the first weather satellite (Tiros) in 1960, and the Russians' operational system (Meteor) followed in 1969. The Chinese government approved the concept of a meteorological satellite in February 1970, but development was impeded by the Cultural Revolution. China set up its first station to receive internationally available meteorological data in 1970 in Beijing, stations being subsequently built in Urumqi and Guangzhou.

The Chinese first expressed a public interest in weather satellites when they revealed the existence of a programme for meteorological satellites called Qi Zing (though this

Feng Yun 1 in preparation.

name never reappeared). The first funds were not allocated to the programme until April 1978. The project has been supported by the Central Meteorological Bureau.

The new satellite was named Feng Yun ('wind and cloud'), and Meng Zhizhong was appointed chief designer. The Feng Yun 1 satellite was hexagonal, 1.76 m tall, 1.4 m wide, weighed 757 kg and had two solar panels spanning 8.6 m. It had a scanning radiometer designed to monitor clouds, water colour, crops, forests and pollution, and pictures were to be transmitted automatically in real time and via a tape recorder. Small nitrogen-powered thrusters were used to ensure the satellite was pointed correctly. The Chinese experienced much difficulty in devising a satisfactory system of gyroscopes to orientate the spacecraft, and they eventually bought some American equipment. The satellite was designed to orbit the Earth 14 times per day.

It was decided to fly Feng Yung in a polar, Sun-synchronous orbit – polar because it crosses the planet on its south-to-north axis, and Sun-synchronous because it follows the same ground track each day and crosses the same point on the Earth's surface at the same time each day. Targets are illuminated at the same Sun angle, which makes it easier to compare weather data from one day to the next. A new launch centre was required for a satellite to enter this type of orbit. Jiuquan was not suitable, since launching northwards would take the departing rocket quickly over Russia, and Xi Chang was too far south. Accordingly, a former missile site near the industrial coal town of Taiyuan was used.

A new launcher was also developed for the Feng Yun satellite – the Long March 4, designed by Sun Jingliang. Whilst the Long March 3 was too powerful for this mission, the Long March 2C had insufficient thrust to send the payload into polar orbit. The Long March 4 was based on the highly reliable Long March 2C, but with an upper stage. Standing 41.9 m tall at liftoff, with a thrust of 300 tonnes, it had a small third stage using conventional fuels.

FIRST POLAR, SUN-SYNCHRONOUS, TAIYUAN, LONG MARCH 4 LAUNCH

The first Chinese weather satellite took a long time to appear, and was eventually launched on 6 September 1988. The Feng Yun 1 launching achieved a number of other milestones: it was the first use of the Long March 4 rocket, the first launch from the new launch site of Taiyuan and the first Chinese satellite to enter polar, Sun-synchronous orbit. Feng Yun 1-1 entered an orbit of 881 × 904 km, inclination 99.12°. It soon sent back pictures of cyclones, rainstorms, sea fogs and mountain snow, and in addition to its meteorological role, it carried other experiments. The main spacecraft carried equipment to detect cosmic rays, protons, alpha particles, and carbon, nitrogen, oxygen and ion particles in the Earth's radiation belts.

Feng Yun 1-1 was less than entirely successful. One of the radiometers failed to work, and the main spacecraft failed after 39 days. Apparently, condensation in the spacecraft had not been fully removed before it left Earth, and this fouled up the sensitive radiometer. Until then, Chinese scientists had been very satisfied with its performance.

FENG YUN 1-2 CARRIES BALLOONS

Feng Yun 1-2 was launched two years later on 3 September 1990. It featured a number of improvements, it was heavier (881 kg), and it carried two additional experiments – two balloons called Qi Qui Weixing 1 and 2. Their purpose was to measure the density of the

Feng Yun 1 lifted by crane in assembly hall.

upper atmosphere between 400 km and 900 km. Measuring 2.5 m and 3 m in diameter, and deployed in similar orbits, they decayed from orbit the following year, one in March, the other in July.

In February 1991, the weather satellite itself appeared to suffer radiation damage, possibly from a solar flare, but after a 50-day struggle, ground control in Xian recovered the satellite fully. There was further radiation damage later in the year, and the data became unusable. However, the Chinese always made it clear that the first two spacecraft were tests before the system would become operational with Feng Yun 1-3 and Feng Yun 1-4. A top priority must be to ensure a working life measured in years, rather than months. Ultimately, the series will be replaced by a new generation, Feng Yun 3, from 1998.

GEOSTATIONARY METEOROLOGICAL SATELLITE

For the Chinese, the next stage was to operate a weather satellite in geostationary orbit. Called Feng Yun 2, this would complement the Feng Yun 1 series. The concept was that Feng Yun 2 would send back constantly scanned pictures of China and the western Pacific from its high vantage point 36,000 km out, while Feng Yun 1 would send back detailed weather maps from its regular 100-minute passes over China from an altitude of around 900 km.

Geostationary meteorological satellites are expensive, requiring a big launcher and high operating standards of the satellite concerned. However, the vantage point of 36,000 km can provide quality weather coverage of large land masses round the clock. The United States operated its first geostationary metsat (Synchronous Meteorological Satellite 1) in 1974, and Japan and Europe (Himawari and Meteosat, respectively) followed in 1977.

Feng Yun 2 in testing.

Feng Yun 2 was a drum-shaped satellite 4.5 m tall, with a diameter of 2.1 m and a weight of 1,380 kg. It carried a multi-channel scanning radiometer made by the Institute for Technical Physics in Shanghai, a cloud coverage information system and data collection translator, and was intended to provide cloud, temperature and wind maps from its vantage point at 105°E. Developed by the Shanghai Aerospace Technology Research Institute of the China Aerospace Corporation and built in the Hauyin machinery plant, it was to operate for several years, and the series was eventually to be replaced by a new generation, the Feng Yun 4.

FUELLING DISASTER IN THE PROCESSING HALL

The Feng Yun 2 series began disastrously. When Feng Yun 2-1 was being loaded with propellant in the processing hall at Xi Chang launch site on 2 April 1994, the satellite exploded, killing one technician and injuring 31 others. The satellite itself, valued at over $75 million, was, of course, a write-off, and it took more than three years to redesign the propellant tank system so as to make sure this type of accident would never happen again.

Some American electronic components were installed in the replacement satellite, which was eventually launched on the troublesome Long March 3 rocket from Xi Chang at 9 p.m. Beijing time on 10 June 1997. Some 23 minutes after launch, the hydrogen-powered third stage fired to send the 1.38-tonne metsat on its way to a permanent position at 105°E, with a scheduled lifetime of three years. It was intended to provide cloud maps, temperatures and wind movements, leading to much improved forecasting. By September it had completed its full range of systems testing and was ready for handing over to the state meteorological administration.

FUTURE REMOTE SENSING PROJECTS: ZIYUAN

China has also worked with Brazil to develop a joint remote sensing satellite on a 70/30 basis. CBERS (China Brazil Earth Resources Satellite) involves the building of two satellites called Ziyuan ('resources'), one in China, and the other in Brazil. Based on the unflown Shi Jian 3 concept, the Ziyuan project has been in the planning stage a long time, having been announced in the mid-1980s. Box-shaped, and with one large 6-m solar panel able to generate 1,100 W, each Ziyuan satellite will enter an orbit at 778 km, inclination 78.5°, similar to that of the Feng Yun 1, and provide detailed images of the Earth in five channels using linear CCDs with a resolution of 20 m. Ziyuan will carry a multispectral infrared scanner (resolution 160–180 m), a high-resolution CCD camera and a large field imager (resolution 258 m). A data collection system will pick up and retransmit information from unmanned inland stations and buoys. The first launch has been set for the 1998 period onwards. Ziyuan, weighing 1,450 kg, will enter a polar orbit repeating every 26 days, having been orbited by a Long March 4B from Taiyuan.

Table 4.5. Chinese weather satellites

Name	Launch date	Notes
Feng Yun 1-1	6 Sep 1988	Failed after 39 days; poor data return.
Feng Yun 1-2	3 Sep 1990	Limited operational life; carried two atmospheric balloons, Qi Qui Weixing 1 and 2.
Feng Yun 2	–	Exploded in preparation, 2 Apr 1994.
Feng Yun 2-1	10 Jun 1997	Positioned over 105° E.

RECOVERABLE SATELLITES: THE NEW FSW 1 SERIES

Chapter 3 examined the introduction of the recoverable satellite programme, Fanhui Shi Weixing (FSW). The FSW 0 series, first flown in 1975, had made nine missions by August 1987, all successful. The FSW 1 series was introduced in September 1987, barely a month after the conclusion of the FSW 0 series. Compared with the FSW 0 series, the '1' series was heavier (2,100 kg) with a greater payload (180 kg), and was able to orbit for up to 10 days (although 8 days was normal). A digital control system was introduced, new gyroscopes were added to help control attitude, new sensors were fitted, the satellite could be re-programmed when in orbit, a control computer was installed and the pressure inside the cabin could be regulated.

The first mission was devoted to microgravity experiments – seeing how algae would grow in orbit, and processing gallium arsenide – but it is not known if the remote sensing package was carried. FSW 1-2 was dual-purpose, carrying both the Chinese remote sensing package and a German protein crystal growth experimental package called Cosima (though it was damaged due to a heavy landing). The German experiment, developed by

FSW 1 launch, 6 October 1992.

Messerschmitt Bolkow Blohm and the German space agency, DFVLR, was intended to find new ways of producing the medical drug interferon from large and pure protein-based crystals. Germany paid China $374,000 for the mission. The Germans were handed back their package the day after landing. Subsequently, a joint company, Euraspace, was formed between Deutsche Aerospace (DASA) and the China Aerospace Corporation, for the development and marketing of Earth observation satellites.

CHINA FLIES ANIMAL PASSENGERS INTO ORBIT

Guinea pigs and plants were carried on FSW 1-3 as part of a microgravity experiment. In doing so, China became the third nation to send animals into orbit and recover them. FSW 1-4 carried a Swedish satellite Freja piggyback into orbit, while the main spacecraft carried Chinese and Japanese microgravity experiments (the latter being a 710° C microgravity furnace). The Chinese experiments involved testing how rice, tomatoes, wheat and asparagus would grow in orbit (apparently, they grew much faster).

In 1988, the Chinese revealed that one of the missions in the FSW series was effectively lost when the second stage vented used propellant over the cabin. Although the mission was carried out perfectly, the photographic results were useless, due to the windows being covered in rocket fuel. The Chinese did not identify the mission in question.

ROGUE SATELLITE ON THE LOOSE

The last mission in the FSW 1 series went wrong – the only one to have done so – when in October 1993 FSW 1-5 failed to return to Earth when commanded to do so. Launched on 8 October 1993, it should have returned to Earth on 16 October after the normal 8-day profile. However, the satellite failed to rotate downward for the return to Earth, and instead its rocket fired in the direction of travel, and sent the FSW into a much higher orbit. As it had been circling the globe at 196 × 251 km, the burn had the effect of making the upper point of the orbit swing far out, to 3,023 km. However, the low point, or perigee, was slightly lower at 181 km, so that the satellite would sooner or later burn up – except that, being designed for re-entry, there was a real possibility that it might survive the fireball of re-entry and crash somewhere on Earth. FSW's orbital inclination of 56° moreover placed at risk anywhere on Earth between 56° N and 56° S – the most inhabited zones of our planet. The media, as ever, warmed to the apocalyptic prospect of a rogue satellite plunging to Earth, as it had earlier with Skylab, Cosmos 954, Cosmos 1402 and Salyut 7. Although tracked carefully by United States space command and other sophisticated radars, the fact that it could fall on most inhabited zones on Earth added to the excitement. Xian control centre kept the FSW under observation, as it had built up its own training by tracking Skylab and Cosmos 1402 much earlier. Experts predicted the size of the crater the satellite could make if it survived re-entry and exploded on impact. Eventually, after several days of bouncing on the upper layers of the atmosphere it came down on that part of the Earth where it was most likely to end up – the sea – and crashed into the South Atlantic on 12 March 1996. For some reason, this particular FSW was emblazoned with a gold-studded medallion of Mao Zedong.

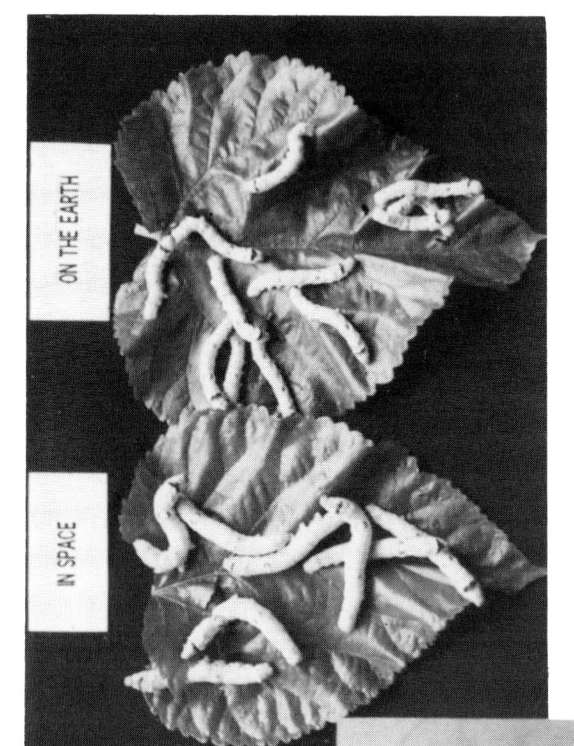

FSW biological tests in Earth orbit.

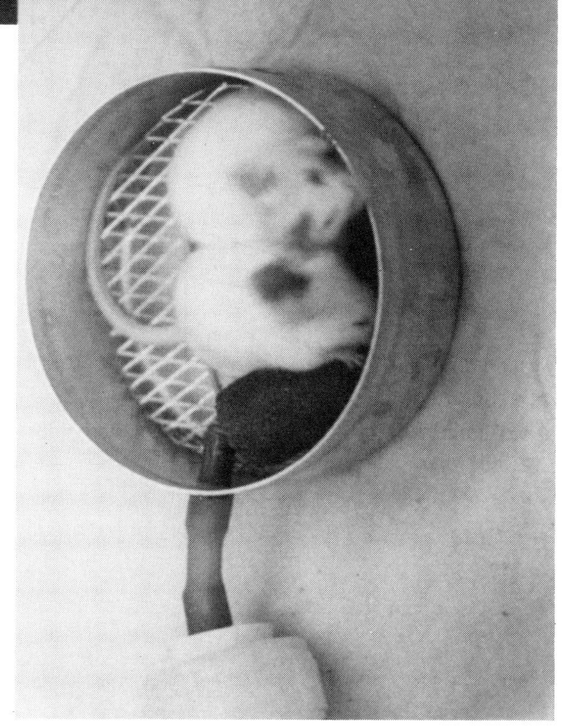

Chinese space animals.

FSW 2 SERIES: MANOEUVRABLE, HEAVIER, 18-DAY PROFILE

The FSW 2 series was introduced in August 1992, even before the FSW 1 series had come to an end. The principal innovation of the FSW 2 series was the ability to manoeuvre in orbit, but there were other improvements. Compared with the FSW 1 series, the '2' model had a greater weight (3,100 kg) and heavier payload (350 kg) and could stay in orbit up to 18 days. Its length was increased by a third to 4.6 m, it had a much larger and more sophisticated service module, part of which was pressurised. FSW 2 carried a more sophisticated attitude control system and an advanced computer. Because of its much-increased size it required a larger launch vehicle, the Long March 2D, an improved version of the Long March 2C.

FSW 2-1 (August 1992) was a dual-purpose mission, with both remote sensing and microgravity experiments (in this case, cadmium, mercury, tellurium and protein crystal growth). The cameras could take 2,000 m of film and had an imaging capability of 10 m. FSW 2-1 carried semi-conductors and 10 protein growth experiments in 48 cells. Crystals were grown in a furnace able to provide a temperature of 813° C. FSW 2-1 used its manoeuvring system to change orbit three times during the mission. The first attempt to re-enter, on the twelfth day of the mission, failed, but after carefully going through the procedures again, ground controllers were successful on the sixteenth day.

FSW 2-2 flew in July 1994 and, like its predecessor, manoeuvred in orbit. It carried an even more exotic cargo of rice, water melon, sesame seeds and more animals. The most significant change, however, was the re-entry manoeuvre. The equipment module remained attached until retro-fire, when the firing of the retro-rockets expelled it into a much higher orbit. Clearly, this had something to do with improved re-entry procedures following the loss of FSW 1-5.

FSW 2-3 carried Japanese microgravity experiments for the Japanese Marubeni Corporation with Waseda University, which involved the development of indium and gallium monocrystals. China also carried its own microgravity materials-processing experiments, as well as a biology package of insect eggs, algae, plant seeds and small animals for the Shanghai Institute for Technical Physics. This was the sixteenth successful recovery out of 17 attempts. For the first time, the information collected by the satellite in orbit was stored on compact disk. The Chinese materials processing experiments concerned the production of monocrystalline silicon, photoconductive fibre with impurities of 10^{-7}, and medicines to prevent haemophilia.

RESULTS FROM THE FSW SERIES

A progress report was issued on the early outcomes of the FSW materials-processing and biology missions to date[27]. By this stage, six such missions had been flown within the series. Tests on alloys, tellurium and gallium arsenide had yielded positive results, crystals having high purity. Rice seeds brought back to Earth and crossed with Earthly grains produced high yield rates, some having 53 per cent more protein. Space-grown yeast offered higher and faster fermentation rates, opening up new prospects for a space beer industry. Algae flourished in orbit.

Around 300 varieties of seed and 51 kinds of plant were carried. Once back from space, with its zero gravity, seeds from the plants grown on board – rice, carrot, wheat, green pepper, tomato, cucumber, maize and soya bean – were planted out by the Institute of Genetics to observe the effects, further note being taken of succeeding generations over the following years. The results varied. Some strains of rice improved from their space experience, while others did not. Some grains grew faster and were fatter, heavier and sturdier. Wheat experiments produced new strains that had short stems and grew fast. One strain of green pepper, called the Weixing 87-2, demonstrated an increased yield of 108 per cent, 38 per cent less vulnerability to disease and an improved vitamin C content of 25 per cent, bearing fruit long after terrestrial peppers had lost their leaves. A fifth generation space tomato had a yield 85 per cent higher than its terrestrial rivals and doubled its resistance to disease. Space-grown cucumbers demonstrated a surprising ability to withstand greenhouse mildew and wilt, female cucumber flowers were observed to flourish in the space environment, and asparagus seeds flown in space have also thrived on Earth[28].

Following the return of FSW 2-3 in November 1996, the Xinhua news agency reported on some of the results of the second series. Among the Earth resources results from the FSW series was a recalculation of the total number of islands off the Chinese coast (5,000, instead of 3,300). The country's farmland had been recalculated at 125.3 million hectares rather than 104.6 million hectares. The Chinese claimed a resolving ability for the FSW cameras of 19 m, which would be at least as good as comparable American Landsat data at the same time. The seven biology packages flown by this time had brought 300 varieties of 51 crops into space, and would make a contribution to the development of more productive varieties of grain.

The following year, further results were announced. The FSW satellites had compiled detailed Earth resource maps of Beijing and its eastern environs, Tianjin and Tangshan. Oil deposits had been discovered in Tarim, chromium and iron deposits in inner Mongolia and coal elsewhere. The FSW satellites had discovered remnants of the Yuan dynasty's ancient city of Yingchang. The Chinese have claimed that the imaging systems have done much to help complete the mapping of China. A new map of China was commissioned in 1949, but only 64 per cent of it had been finished by 1982. However, 600 FSW pictures were able to finish the job in a matter of months. FSW satellites have been used to prepare geological survey maps, identify the optimum routes for railway lines, and track the patterns of silting in the Huang (Yellow), Luan and Hai rivers. Images have tracked the path of the Great Wall across northern China and found the old walls of the Chengde summer palace. They even uncovered buildings erected by the first Yuan emperor, Kublai Khan, for his daughter, Princess Luguo Dachang, in 1270. They have tracked water and air pollution, observed soil erosion and identified geological fault lines. The FSW satellites located seven new mineral deposits for the Capital Iron and Steel Works, Beijing, found goldfields in Mongolia and oil and natural gas in the Yellow River delta and offshore. Data from the FSW and Feng Yun series, combined with information from the American Landsat and the French SPOT satellites, provided a worrying picture of desertification in Qinghai in the north-west. Dynamic changes were taking place, according to the satellite data: dunes had advanced, grassland was damaged, and water resources had been misused. Elsewhere, soil erosion had been noted. Positively, the rate of afforestation had been assessed and was seen to be increasing. The use of windbreak forests in northern China had already regenerated the ecology of the area. Earth resources satellites carefully tracked the evolution, speed and impact of the Yellow River, and as a result, timely warnings about floods were announced during the inundations in 1991, minimising damage. Satellite tracking of the 1987 forest fires in Xinanlang enabled firefighters to save up to 10 per cent of the forests from further damage.

At one stage – when it was flying French and then German payloads – the series appeared to hold out the prospect of being a serious foreign exchange earner for China. Both the United States and Europe lacked dedicated spacecraft able to fly microgravity experiments. The only countries with such a capability were Russia (the Foton series) and China (FSW). However, after the Tiananmen Square massacre, the Germans moved their payloads to comparable Russian spacecraft.

As military reconnaissance satellites – its probable original role – the FSW series had definite limits. FSW film was sent back in recoverable cabins, and military analysts had to wait until the film was developed before they could analyse points of interest. In 1977, the United States introduced downlink imaging: satellites could transmit photographs to the ground directly from orbit. Analysts would have their intelligence information immediately, and could even command the spy satellites to adjust course to investigate new and more interesting targets later in the mission. (With their fifth generation of photoreconnaissance satellites, the Soviet Union had a digital imaging capacity from 1982). However, it seems that China did not develop the series for military photoreconnaissance, adapting the series instead for civilian Earth resources studies, microgravity experiments and for commerce.

Table 4.6. FSW overall programme summary

Series	Launch dates	Notes
Fanhui Shi Weixing	1975–1986	10 attempts, 1 launch failure (1974).
Fanhui Shi Weixing	1987–1993	5 attempts, one recovery failure.
Fanhui Shi Weixing	1992–	3 attempts, no failures.

ASSESSMENT AND CONCLUSIONS

The development of the Long March 3 series, the use of hydrogen fuels, and the building of geostationary communications satellites were ambitious steps for the Chinese space programme at an early stage. The scale of effort involved was considerable. Careful development, rigorous testing and quality control lead to early success, and by the early 1990s, Chinese launchers had become a part of the world launcher market and had established an important niche in orbiting Asian, Australian and some American payloads. But the series of failures which beset the programme in the mid-1990s must have been deeply disappointing. The programmes of the big space powers – Russia, the United States and Europe – have all known frustrating periods of failure and difficulty, but with new orders in the late 1990s, it is probable that confidence will return to the customers of Chinese rockets. The question of the reliability of Chinese rockets is explored in more detail in chapter 6.

Table 4.7. Programme summary: FSW 1 series

Name	Launch date	Notes
Fanhui Shi Weixing 1-1	9 Sep 1987	Gallium arsenide crystals; algae growth experiments; consignment from Matra.
Fanhui Shi Weixing 1-2	5 Aug 1988	Three German experiments.
Fanhui Shi Weixing 1-3	5 Oct 1990	Guinea pigs – first Chinese animals in orbit.
Fanhui Shi Weixing 2-1	9 Aug 1992	Semi-conductor and protein crystal growth experiments.
Fanhui Shi Weixing 1-4	6 Oct 1992	Gallium arsenide, rice, wheat, asparagus, algae.
Fanhui Shi Weixing 1-5	8 Oct 1993	Recovery failed; crashed to Earth, March 1996.

Table 4.8. Programme summary: FSW 2 series

Name	Launch date	Notes
Fanhui Shi Weixing 2-1	9 Aug 1992	Recovered after 16 days.
Fanhui Shi Weixing 2-2	3 July 1994	Recovered after 13 days; new re-entry procedure following FSW-1-5 failure.
Fanhui Shi Weixing 2-3	20 Oct 1996	Recovered after 15 days; Japanese experiments.

The FSW series – now in its third decade – was no mean achievement. Designed before China had even put an Earth satellite into orbit, the FSW satellite was a demanding project for a space programme at such an early stage of development. The series has been ridiculed in the West, where it was misrepresented as low-tech, using a tourist camera with regular film and even equipped with a heat shield made out of wooden (oak) planks[29]. In fact, the development of recoverable spacecraft is a challenging engineering task, one which none of the other minor space powers has yet undertaken. The FSW series has shown a high level of reliability, with only one failure out of 17 missions.

The success of the FSW series in returning Earth resources data probably encouraged the Chinese in the development of meteorological satellites. Although the Feng Yun 1 series had operational problems, the quality of data and film returned has been very promising. With the Feng Yun 2 series of meteorological geostationary satellites, the Chinese again chose an ambitious option, but one guaranteed to present a maximum return from investment.

5

Making it possible: China's space industry

This chapter goes behind the scenes of the public face of the Chinese space programme, such as its launchers and satellites. In terms of size, the Chinese space programme may be about the sixth largest in the world, but would probably be the largest of the developing countries. This chapter provides an overview of the Chinese space programme in a comparative international perspective, examining its launch rate compared with other countries, the main areas of space activity, the budget and scale of the programme and the rhythm of launches. It also examines the role of the chief designers of the Chinese space programme, how they were recruited and how they related to the political leadership of China. The various design bureaux are described, from the rival bodies of the 1960s to the consolidated space industry of today. The main lines of command in the Chinese space programme are described, with the respective roles of the Chinese National Space Administration, the China Aerospace Corporation and the various design academies. There is a description of the main ground testing infrastructure, whose aim is to produce space hardware to the highest possible standard, and other important parts of the programme, such as the Institute for Aviation and Space Medicine and the Academy of Sciences are reviewed. China has an extensive system for tracking spacecraft, starting with the main mission control centre in Xian and including three communications ships at sea. The benefits of the space programme to the Chinese economy are reviewed, and the chapter concludes by taking a brief look at China's international links and contacts.

THE CHINESE SPACE PROGRAMME IN COMPARATIVE INTERNATIONAL PERSPECTIVE

If we define a space power as a country or block able to put its own satellite into orbit, then the world has nine of them: Russia, the United States, France, Britain, Europe, China, Japan, India and Israel (Britain cancelled its launcher programme, but permitted one launch attempt to be made; in the event it was successful). These distinctions are in some ways artificial, for many other countries have space programmes, but they use other countries' launchers. The differences are becoming artificial because of the arrival in the 1990s of a range of commercial companies providing rocket services. National boundaries have become less important: the Sea Launch project, to send rockets into orbit from the mid-

Pacific, is developed by Boeing, owned by Russian and American companies and uses Ukrainian rockets fired from a Norwegian oil rig serviced by a command ship built in Scotland. Nevertheless, it is important and valuable to set the Chinese space programme in a comparable international perspective. China accounts for a very small proportion (1.13 per cent) of world space launches, but the proportions are much higher if one takes the two superpowers, Russia and the United States, out of the equation. Of the 196 launches of the minor powers, China then accounts for almost 22 per cent of launches. Table 5.1 lists the number of launches by the different space faring nations.

Table 5.1. Number of successful space launches, 1957–96

Soviet Union/Russia	2,519
United States	1,088
Europe	84
Japan	49
China	43
India	7
Israel	3
France	10
	3,803

Adapted from Harris, R.P., *Living and working in space*, Wiley–Praxis, 1996 (2nd edn.)

In terms of the level of development of the respective space programmes, the following factors may be taken into account. Only the Soviet Union, now Russia, and the United States have the ability to launch manned spacecraft, though Europe and Japan have toyed with spaceplane designs. Many nationalities have now been into space, but as guests of the two superpowers. Looking at deep space missions (the Moon, Venus, Mars and beyond), four of the space powers – the United States, Russia, Europe and Japan – have launched deep space missions, but China has not. Turning to geostationary orbit, only five

Table 5.2. Number of world space launches, 1996

United States	33
Russia	26
Europe	10
China	3
Japan	1
India	1
	74

countries or blocks have launchers able to reach 24-hour orbit: the United States, Russia, Europe, China and Japan. With Russian help on the upper stage, India may have the capacity from 1998.

Placing China in the context of current world space activity, Table 5.2 lists the number of Chinese space launches compared with the rest of the world. Thus it may be seen that in 1996 China accounted for 4 per cent of world launches, or 20 per cent of those of the minor space powers.

CHINA'S SPACE BUDGET

Estimating China's space budget has always been problematical. As in the Soviet command economy, financial transfers between organisations are often set at notional amounts. Furthermore, important functions in the space programme are performed with military help (for example, the rocket troops, search and recovery operations). Another consideration is that labour costs in China are exceptionally low. From the period of the four modernisations, self-accounting has been much more in evidence; indeed, the Chinese claim that they developed the Long March 2E entirely on their own company resources, without any state help. The Chinese themselves estimate that government support for space activities is worth 1.45 billion yuan annually – about £154 million. However, this may simply be the research and development figure, for it is known to exclude launcher operations. Several authoritative western estimates have been made, some close to each other. These are in the range of $1.35 billion (America's *Aviation Week & Space Technology*) to $1.38 billion (Britain's *Flight International*), or between £831 million and £850 million. This places China as the fifth largest space spender in the world. (The Russian figure is problematical, as it understates the programme's huge capital assets.)

Table 5.3. Estimated World space budgets, 1997

United States	£11.02 bn
Europe	£2.27 bn
France	£1.79 bn
Japan	£1.66 bn
China	£0.85 bn
Germany	£0.71 bn
Russia	£0.54 bn
India	£0.25 bn

Adapted from *Air and Cosmos*

There are no absolutely clear figures available for the numbers of people working in the Chinese space programme, and the best Western estimates give a figure of 200,000 directly involved in the space industry. Of these, 100,000 are technical workers, drawn from light industry, the army's technical ranks and the polytechnical schools, and about 10,000 are graduate research engineers working in 460 institutes connected to the space pro-

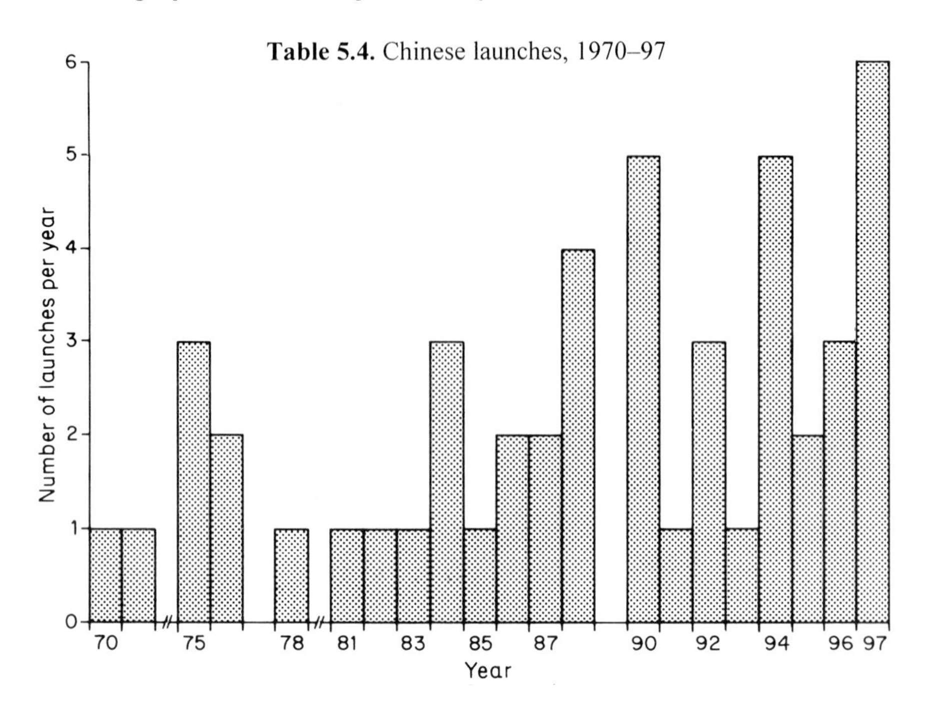

Table 5.4. Chinese launches, 1970–97

gramme. The Chinese space programme has been able to choose the top graduates coming out of engineering schools, and also to attract the country's most talented scientists.

THE RHYTHM OF CHINA'S SPACE PROGRAMME

There are several approaches to analysing the rhythm and characteristics of the Chinese space programme. Table 5.4 shows the total number of successful launches made by China, and the following are the basic statistics (end of 1997):

Number of launches in which payloads reached orbit: 49
Number of satellites put in orbit: 58, of which 14 (including the Aussat model) have been commercial satellites for foreign customers.

There are several ways of calculating such tables. This table (and the book as a whole) categorises a launch as one in which a payload reaches low Earth orbit intact, which produces a launch rate of 1.8 per year. As can be seen, the launch rates of the Chinese space programme are low and have never exceeded six in any given year, while in some years (1989, for example) there have been no launches at all.

By 1998, China had launched 58 satellites, the categories of which are shown in Table 5.5. The recoverable satellite series is the largest single element of the programme, accounting for 29 per cent of missions, followed by foreign commercial communications satellites (23 per cent).

THE CHIEF DESIGNERS AND THE ENGINEERS

A rocket programme is a reflection of the quality of its leadership. It is doubtful if the American space programme could have conquered the Moon without the genius of rocke-

Table 5.5. Classification of satellite types

Recoverable	17	29
Geostationary (domestic)	10	17
Scientific[†]	8	14
Meteorological	3	5
JSSW series	3	5
Commercial	13	23
Demonstration	4	7
	58	100%

[†] Dong Fang Hong is included as scientific; Aussat model, KF 1, Iridium tests are classified as demonstration. This table does not include unsuccessful launches in which the payload failed to enter Earth orbit. Percentages are rounded.

teer von Braun or Apollo designer Maxime Faget. Similarly, the Soviet space programme attracted the genius of Sergei Korolev, Valentin Glushko, Vladimir Chelomei and others. In China, the scientific community faced particular political problems.

In the mid-1950s Nie Rongzhen was charged by the Central Committee with the recruitment of the men and women who would build the Chinese missile space programme. His first efforts were to persuade the many Chinese people who had gone to study abroad in the 1930s, 1940s and early 1950s to return home. Aided by Zhou Enlai as Foreign Minister and then as Prime Minister, he appears to have been spectacularly successful: the biographies of China's space designers look like a graduation list from Massachusetts Institute of Technology and California Institute of Technology. Beside Tsien and many other rocketeers already mentioned, Nie Rongzhen brought home materials specialist Yao Tongbin, aerodynamics expert Zhuang Fenggan, automatic control designer Yang Jiachi and electronics expert Huang Chang. Zhou Enlai went to some lengths to ensure that the communist party welcome them home and not treat them with suspicion because they had been born in what was termed 'the old society' or were not communists. His approach was to come under severe strain during the anti-rightist struggle of 1957, the campaign against experts in 1958, and during the Cultural Revolution of 1966–76. Zhou Enlai, Nie Rongzhen and Zhang Aiping were put to considerable efforts to defend the scientists from those who saw intellectuals as likely class enemies. Nie Rongzhen felt obliged to explain that 'real socialism' involved scientific achievement more than political correctness, whilst Zhou Enlai weighed in with the axiom that 'if we serve scientists, the scientists will serve socialism'.

During the great leap forward, many scientists (though not, apparently, Tsien) made clear their contempt for the principles underlying the campaign against the four pests, such as sparrows. During the Cultural Revolution, there was strong anti-intellectual sentiment. Long tracts were written on the class status of the intellectual in building communism: indeed, it was one of many themes that obsessed the Red Guards. Practical compromises were made with the revolutionaries: they mainly took the form of extra ideological and political classes for the intellectuals, so that, in the pithy expression of a slogan of the period, they could be both 'expert' and 'red' at the same time. During rectification, ideology class times were cut back to no more than one sixth of working time. In 1978, Deng Xiaoping effectively declared this debate to be closed by offering a theoretical and strategic clarification of the problem: he calmly announced that China's intellectuals were now an integral part of the working class and that was the end of the matter. Job titles were restored and the pre-1966 system of ranks and promotion resumed.

In addition to recruitment from abroad, Nie Rongzhen drafted university graduates into the Fifth Academy. Using the powers which are available in a command economy, graduates were requisitioned from the military engineering academy: a state order was issued, called simply *Notice concerning the transfer of university graduates to the Fifth Academy*, and quotas were set. In 1960, 1,000 senior military officers – some of whom were grisled veterans from Mao's army – were drafted into the space programme. Most knew nothing of science, but they studied furiously and learned to get on with the intellectuals (the scientists), and some even became friends[30]. Nie Rongzhen's drafting of graduates, with promotion by merit and fast-track promotion of talented workers, was politically contentious. The policy, officially termed 'the cream of the crop', was challenged as inegalitarian and was suspended for the Cultural Revolution. In a further move, he sent students to study abroad in the Soviet Union and eastern Europe (though this did not last) and commandeered the services of those who had studied there earlier.

FIRST GROUP OF CHIEF DESIGNERS

The first group of chief designers was appointed when the Fifth Academy was set up (though for some reason not formally established as a system until May 1962). Their leaders are shown in Table 5.6. Like their counterparts in the United States and the Soviet Union, most of these men were well into their forties, some in their fifties, when they led the Chinese space programme in its formative years. They have since retired, though in

Table 5.6. First group of chief designers and their responsibilities, 1958

President	Tsien Hsue-shen
General programme office	Ren Xinmin
Aerodynamics	Zhuang Fenggan
Structures	Tu Shoue
Engines	Liang Shoupan
Propellants	Li Naiji
Control systems	Liang Sili
Controls components	Zhu Juingren
Radio systems	Feng Shizhang
Computers	Zhu Zheng
Technical physics	Wu Deyu

keeping with the Chinese tradition of longevity, many have lived to old age. They had a powerful sense of mission, and enormous dedication and will to overcome difficulty. For those who had left (or been forced to leave) the United States, the change in their living conditions and remuneration must have been dramatic.

From the 1970s, a new generation of leadership emerged. Many were the young graduates transferred into the space industry in the 1960s. They came principally from Harbin Military Engineering College, the National Defence Science and Technology University, Beijing University, Qinghu University, Beijing Aerospace College, Beijing Industrial College, Harbin Industrial College and Northwest China Industrial College. The first of them began to move into leadership positions during the period of rectification: people like Song Jian (promoted to chairman of the state science and technology commission) and Li Baijong (promoted to president of CALT).

DESIGN BUREAUX AND ORGANISATION

The quality of management and organisation of a space programme is one of the keys to its successes. One of the least understood achievements of the Apollo space programme which put American astronauts on the Moon was the way in which clear objectives and lines of command ensured speedy, effective decision-making. By contrast, among the reasons why the Russians lost the race to the Moon were confused objectives, unclear lines of command and rivalry between design offices which Soviet politicians and bureaucrats were unable to tame. Likewise, the Chinese space programme spent most of its earlier period in some organisational turmoil. The 1960s and 1970s saw rivalry between the institutes based in Shanghai and Beijing.

From its foundation in 1956, the Chinese space programme was coordinated by the Fifth Academy, which from 1964 became the Seventh Ministry. In neither case was the true nature of the respective body revealed outside the bureaucracy itself. Not until May

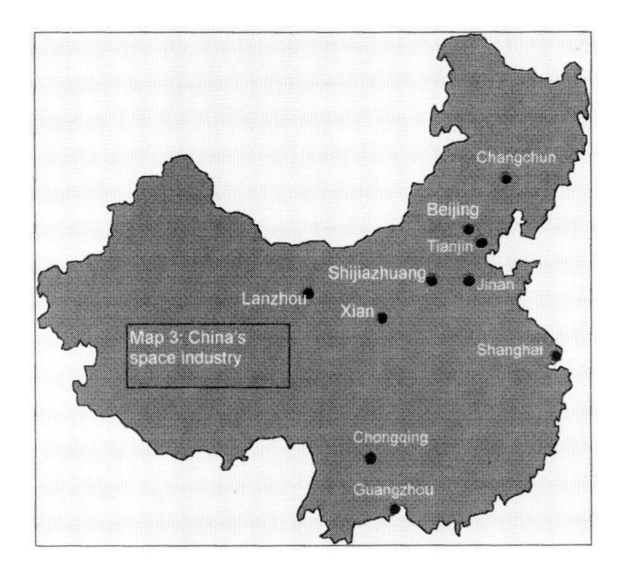

Map of China's space industry.

1982, when the Space Ministry was formally established with Zhung Jun as its first minister, did the Chinese space programme have an identifiable public identity.

Who's in charge? Ruling bodies of the Chinese space programme

8 October 1956	Fifth Academy
23 November 1964	Seventh Ministry
9 April 1982	Ministry of the Space Industry
1993	Chinese National Space Administration

In the United States, rockets and satellites are built by the large aerospace corporations under contract to NASA. In Russia, space programmes are developed by design bureaux, or OKBs, which are awarded responsibility for particular missions and areas of development. The broad lines of the Chinese organisation were laid down from the late 1950s and follow a pattern broadly in line with the Soviet system[31].

CHINESE NATIONAL SPACE ADMINISTRATION (CNSA)

The Chinese space effort now comes under the Chinese National Space Administration, established in 1993, the direct equivalent of NASA in the United States or the Russian Space Agency (RKA), and the main policy-making body. The CNSA is responsible to the Prime Minister, government and party, and is an executive agency which replaced the space functions of the old Ministry of Aeronautics and Astronautics. The Director is Liu Jiyuan.

There is also, within the government, a Space Leading Group of the state council. Set up in 1991, its primary purpose is to coordinate relationships with foreign governments, especially in China's efforts to attract foreign contracts, though it has a broader role in policy-making. Its members include the Prime Minister, the chairman of the state Commission for Science, Technology and the National defence, the vice-chairman of the state Committee for Science and Technology, the minister of the aerospace industry, the vice-minister of foreign affairs and the vice-chairman of the state Committee for Central Planning.

The work of the CNSA is required to conform to the work of the State Commission for Science and Technology for National Defence, which has a policy-making and directional role in the organisation of China's scientific, research and industrial progress. (The Commission for Science and Technology for the National Defence is broadly similar to the Military–Industrial Commission of the former Soviet Union.) The Commission is directly responsible for the three main launch centres.

CHINA AEROSPACE CORPORATION

The Chinese space industry is brought together by the China Aerospace Corporation (CASC), currently directed by Liu Jiyuan. Founded in June 1993, it has 270,000 employees or people working in sub-contracted industries, and has overall authority for the main industrial groups concerned with spaceflight. These are China Jiangman Space Industry, Sichuan Sanjiang Space Group, Shaanxi Lingman Machinery Co., and the export agency the China Great Wall Industry Corporation (CGWIC). CASC is divided into seven academies:

1. Chinese Academy of Launcher Technology (CALT), in Nan Yuan, Beijing
2. Chinese Academy of Mechanical and Electrical Engineering (CCF), in Beijing
3. Chinese Electromechanic Technology Academy (CHETA), in Haiying
4. Chinese Academy for Solid Rocket Motors (ARMT), in Xian
5. Chinese Academy of Space Technology (CAST), in Beijing
6. Shanghai Academy of Space Technology (SAST), in Shanghai
7. Chinese Academy of Space Electronic Technology (CASET), in Beijing.

Some of these are of great importance, the most eminent being CALT, CAST and SAST.

Academy 1: Chinese Academy of Launcher Technology

The Chinese Academy of Launcher Technology (CALT), originally called the Beijing Wan Yuan Industry Corporation (established in 1957), is located in Nan Yuan, 50 km south of Beijing, and was once a forbidden city. CALT employs 27,000 people: of these, 100 are professors, 2,000 are senior engineers, 7,000 are engineers and 10,000 are technicians in 13 institutes and six factories, mainly in Beijing and Shanghai. Its current president is Li Jianzhong.

CALT has overall responsibility for the Long March 1, 2C, 2E, and 3, Dong Feng 5, and the hydrogen-powered upper stages of Long March rockets. In practice, Long March 2, 3 and 4 rockets are assembled in Shanghai in a horizontal position on a factory floor. Four complete rocket assemblies may be handled there at any one time. The stages are then transported by rail to the appropriate launch site. The Chinese Academy of Launcher Technology and the Shanghai plant between them have the capacity to manufacture up to 16 rockets a year, but demand has never reached such a level of possible activity.

CALT has its own railway terminus, linked to the national railway grid, Shanghai and, of course, the launch sites of Jiuquan, Xi Chang and Taiyuan. Adjoining the institute are residential blocks for the many scientists, engineers and technicians who work there. The research units test out new materials, parts and components. CALT is responsible for a 6,000 m^2 static test hall, a 50-m tall vibration test tower, engine test stands and a moored test stand. These are vital facilities for a space industry and are described in more detail later.

CHINESE SOCIETY OF ASTRONAUTICS

The Chinese Academy of Launcher Technology includes an amateur Chinese Society of Astronautics (CSA) which attempts to bring together engineers, scientists, amateurs and enthusiasts of space flight. It is the body affiliated to the International Astronautical Federation, though in the best traditions of science there is a rival Chinese Society of Aeronautics and Astronautics.

Academy 5: Chinese Academy of Space Technology (CAST)

Formed on 20 February 1968 by Zhou Enlai, with Tsien Hsue-shen as its first head, the Chinese Academy of Space Technology, Beijing, consolidated the range of bodies then engaged in the most advanced aspects of space research. It is the primary body which designs and manufactures scientific and applications satellites. It has 10,000 workers in 10 institutes and three factories. CAST has lead responsibility for the FSW programme and sounding rockets. The president is Xu Fuxiang, with Zhu Yilin the secretary general of its science and technology commission.

One of its principal bodies is the Institute of Control Engineering in Zhongguancun, Beijing, set up in 1956, which has played a key role in satellite design and construction. The director in the 1970s was Tu Shancheng, though it has since been led by Zhou Gangrui, an expert in the control of satellites and the development of sensors. He and his colleagues had to work by night during the Cultural Revolution. China's spacecraft are brought together at the Satellite Assembly Plant in Beijing, which until 1967 had been a precision instrument factory.

Academy 6: Shanghai Academy of Space Technology (SAST)

The Shanghai Academy of Space Technology (SAST), sometimes loosely referred to as 'the Shanghai bureau', was set up in 1961 by Mao Zedong as part of an attempt to develop technological engineering in the industrial centre of Shanghai. Located in Minhang, Shanghai, SAST had overall responsibility for the Feng Bao, the JSSW military satellite programme, and now the Long March 2D and 4. SAST makes the guidance and attitude control systems of the Long March 3 and builds the actual rocket in the Xinxin machinery factory, a converted tobacco hall. SAST has 10 institutes and 12 factories, employing 24,000 people, including 6,000 engineers. The president is Zhang Wenzhong.

Shanghai Institute for Satellite Engineering (HAUYIN)

The Shanghai Institute for Satellite Engineering builds the Feng Yun metsats and previously built the JSSW series of satellites. The institute may have the best technical facilities in Shanghai, with three vacuum chambers and a centrifuge. This institute was part of SAST until 1993, when it became independent. The director is Prof. Meng Zhizhong. Imaging systems are made by the State Meteorology Administration and the Shanghai Institute for Technical Physics (established in 1958).

The other academies: ARMT, CASET and CHETA

The third academy of CASC, the Chinese Electromechanic Technology Academy (CHETA) was set up in 1961 and now employs 15,000 people in 10 institutes and two factories building cruise missiles. The fourth academy of CASC, ARMT (established in 1962) makes solid rocket motors in Xian city. The president is Ye Dingyou. ARMT employs 1,200 people making solid rocket motors for the EPKM kick stage and retro-rockets for the FSW recoverable cabins. The seventh academy of CASC, the Chinese Academy of Space Electronic Technology (CASET) is in Beijing. The president is Tao Jiaqu. CASET has 10,000 people in nine institutes, two factories and five technical centres.

Great Wall Industry Corporation

The Great Wall Industry Corporation is the promotional agency at home and abroad for the China Aerospace Corporation (CASC). Its offices in China may be found in Beijing

(Haidian), Chongqing (Sichuan), Guangzhou, Shanghai and Jinan; abroad, they may be found in California, Washington DC and Munich, Germany. The Great Wall is a multi-product promotional agency, its current portfolio including – besides space rockets – bicycles, beer, safes, home-made ice-cream machines and electric fans. Its president was Zhang Tong, but, on a visit to the European space base in French Guyana in March 1997, while posing for a photograph beside the shore, he was tragically swept away by a freak wave and drowned. His successor was Zhang Xinxia.

THE TESTING INFRASTRUCTURE

Testing and quality control have been an essential element of the Chinese space programme, and considerable resources have been devoted to building up the machinery of testing and verification. Other parts of the space industry are scattered throughout China in a range of facilities, institutes, bodies and companies, large and small. All work closely with the China Aerospace Corporation (CASC) and its relevant academies.

STATIC TEST HALL

CALT's static test hall was, when built in 1963, the largest building in China. It took eight months to construct and involved the driving of 1,300 piles – some as long as 10 m – and two pourings of more than 5,000 m^3 of seamless concrete. Entire rockets can be tested there. For the development of the communications satellite, a large vertical dynamic equilibrium machine was developed. Construction of the machine began in 1976, and it was operational five years later.

VIBRATION TESTS

Vibration tests are essential if the strains put on a rocket during the ascent are to be anticipated. No amount of theory prepared the rocketeers of the 1950s and 1960s for the stresses imposed on rockets as they were shaken by the vibration of their own engines and the strains of passing through the dense layers of the atmosphere. CALT's vibration test tower is unprepossessing, looking like a shabby yellow and orange brick grain mill, but is able to test all the probable stresses an ascending rocket is likely to experience. Built in 1963, the vibration tower is 50 m high with 13 floors and 11 working levels. Entire rockets are hoisted into place on the stand, gripped by bearing rails on the floors and then shaken to exhaustion by 20-tonne hydraulic vibration platforms.

ENGINE TESTS

An essential element of any rocket development programme is a comprehensive facility for engine testing. The main site for testing rocket engines is the Beijing Rocket Engine Testing station, part of CALT, located 35 km from Beijing (the Russian equivalent is in Sergeyev Posad). The director is Xia Zhao Xong. The first rocket test stands were completed in 1964 under the direction of Wang Zhiren, who designed them to run up to four engines at a time and simulate high-altitude tests. The largest engine test stand, completed

Vibration test tower.

in 1969, is 59 m high and has a cooling system which draws on a tank holding 3,000 tonnes of water which cools the engines with 35,370 nozzles, dousing the rocket with 7.9 tonnes of water per second. The engine tests are fed by eight tanks which hold propellants and oxidiser.

In the early 1980s, the station employed 750 people, including 250 engineers and technicians. It is guarded by a strong military presence. The first stand was built for initial rocket tests in the 1950s, and stands 2 and 3 in anticipation of the Dong Feng rocket developments in the 1960s. Stand 1 now handles attitude thrusters. Stand 2 was subsequently modified to test the liquid hydrogen third stage of the Long March. Stand 4, built in 1963, is designed for all-up testing of a complete launcher. Stand 5 is available for horizontal tests of rocket engines and hydrazine thrusters. The centre has a range of facilities available to supply the full range of fuels required for long engine tests.

The moored test stand is 30 m tall and has a 33-m flame trench. This is designed for all-up, full-power tests of entire rocket systems, not just individual engines. For a rocket in development, this is the last phase of static testing. Rockets under test are held between two 33-m high concrete pillars set in reinforced concrete 23 m deep.

Engine test stand.

OTHER SPECIALISED FACILITIES

Wind tunnels are a speciality of the Aerodynamics Research Institute, Beijing. The first were built in China in the defence ministry in 1959, and they played an important role in determining the air flow and pressure on climbing rockets and the successful execution of staging.

To test against leakages in satellites, the Beijing Satellite General Assembly Plant developed a highly sensitive leakage detector using krypton-85, able to pick up a leakage of 50 μm (half the width of a human hair). For the Long March 3 third stage, the Lanzhou Physics Institute developed a helium mass spectrum leakage detector.

One of the strangest test facilities is that used for testing out satellite radio systems. In order to eliminate the possibility of interference, the main requirement of the test hall is that it have no metallic components. Accordingly, it is made entirely of glued red pine wood.

Thermal vacuum tests are essential if a satellite is to withstand the intense cold of the space night and the great heat of the space day, complicated by vacuum (10^{-13} torr at geostationary altitude). A series of vacuum chambers, able to simulate up to 10^{-7} torr, were built in Beijing and Shanghai in the 1960s. The largest vacuum chamber for testing spacecraft before flight, the KM_4, may be found in the Environmental Simulation Engineering Test Station in Beijing. Spacecraft are lowered by crane into the 7 m diameter, 12 m high chamber where they may be alternately frozen, heated, shaken and exposed to a vacuum. Supercooled helium is the chief agent for freezing the chamber, while pumps are used to suck the air out. There is another vacuum chamber in the Huayin Machinery Plant in Shanghai.

INSTITUTE FOR AVIATION AND SPACE MEDICINE

The Institute for Aviation and Space Medicine (many variations of this name appear) in Beijing, is the main centre for the preparation for manned spaceflight and life-support systems, and was established in the late 1960s. The one person mainly responsible for the development of aviation and space medicine in China is Cai Qiao (b. 1897). From Jieyang in Gungdong, he studied psychology and then medicine in California, Chicago and subsequently in London and Frankfurt. He received senior appointments soon after 1949, and has directed research into aviation and space medicine from the very beginning, designing test centres and carrying out research into survival at altitude. He has been published in six major books and over a hundred papers, his main texts being the *ABC of aviation medicine*.

In 1984, the institute developed a Gemini-class space suit. In the early 1980s, five men spent a month in a high-altitude chamber at a pressure of 7 psi. In spring 1997, there were reports that an underwater tank had been built in Shanghai to train cosmonauts in simulated weightlessness, though underwater tanks are normally associated with space walks. The president of the institute is Min Guirong.

There are two large centrifuges in China, the most advanced of which is the large centrifuge operated by the Institute for Aviation and Space Medicine which has an arm of 12 m, is computer-controlled and can reach a maximum acceleration of 25 g. It is used to test the effects of high gravities on spacecraft, during ascent or descent, but can also be used for training astronauts. The Shanghai Research Institute of Satellite Engineering has a 15-m long centrifuge, the biggest in Asia, which can achieve 17 g.

CHINESE ACADEMY OF SCIENCES (CAS)

As in the Soviet (now Russian) space programme, the Academy of Sciences makes an important contribution by the provision of advice, personnel and facilities, and a space committee was set up in the academy in 1963. The Academy of Sciences has five departments: mathematics, physics and chemistry; technology; biology; Earth sciences; and philosophical and social sciences. The first two are of the greater importance for the space programme, the former having institutes of mathematics (Beijing), physics (Beijing), dynamics (Beijing), chemistry (Beijing), applied chemistry (Changchun) and observatories in Beijing and Nanking. A Space Science and Technology Centre was formed in the late 1980s with departments of space science, space technology, space programmes, remote sensing and ground stations. One of the institutes of the Chinese Academy of Sciences is the Xian Institute of Radio Technology (XIRT), established in 1956 and transferred to CAST in 1968. XIRT has played a leading role in developing the electronic capabilities

Table 5.7. Key institutes and bodies in the Chinese space programme

Institute	Area
Shanghai Electronic Equipment Factory	Spacecraft electronics
Shansi Electronic Equipment Factory	Spacecraft electronics
Tung Fang Scientific Instrument Plant, Beijing	Spacecraft testing
Lanzhou Space Research Institute	Low temperature and vacuum testing
Xinyue Mechanical Electronics Plant, Shanghai	Gimballing systems; precision instruments
Scientific Instrument Factory and Institute of Technical Physics, Shanghai	Sensors
Shen Si Institute of Micro-Electronics	Scientific instruments for spacecraft
Institute of Space Physics, Xian	Computer software
Institute of Radio Technology, Xian	Radio communications systems
Institute of Physics, Lanzhou	Ion engines
Shaanxi Liquid Rocket Engine Co.	Rocket engines
Xian Institute of Space Technology (XIRT)	Charge coupled devices
Beijing Institute of Control Engineering (BICE)	Sensors and antennas
South-West Institute of Electronics Technology 54th Institute, Shijiazhuang	Tracking systems
Institute of Power Sources, Tianjin	Solar panels
North-West Institute for Electronics, Xian	Antennas and ground station equipment

of Chinese satellites. The department of technology includes institutes of electronics (Beijing), cybernetics (Beijing), metallurgy (Shanghai), new metals (Shenyang) and optics (Changchun). Two Academy of Sciences institutes in Beijing are of particular importance: the Centre for Space Research and Applied Sciences, and the Laboratory for the Detection of Microwaves, Technology and Information. In 1994, the academy broke the first ground of the new Beijing Centre for Payloads and Applications.

MISSION CONTROL: XIAN

Proper control centres are essential elements in national space programmes. The Americans, for example, use facilities in Houston, Texas, for manned missions, and the Jet Propulsion Laboratory in Pasadena, California, for deep space missions. China first began space tracking soon after the Soviet Union launched Sputnik 1 in 1957. Telescopes and radio receivers were installed in Beijing, Nanjing, Guangzhou, Wuhan and Shaanxi. In

Chinese tracking station.

1958 an artificial satellite motion theory laboratory was founded by Shang Yushe of the Purple Mountain observatory (Nanjing), where he collected data from observations and predicted future satellite passes. In early 1965 Nie Rongzhen and Zhou Enlai conferred on establishing a ground tracking network in anticipation of China's own first satellite, and in 1967 the existing network of satellite observation stations made a series of radio observations of foreign satellites to refine their observational techniques. Since then, the tracking network was established as an organisational command under the State Commission for Science, Technology and the National Defence. It has 20,000 workers, including 5,000 engineers.

China's tracking system

Mission control: Xian
Tracking sites: Weinan, Shaanxi province, near Xian
Min Xi (Fujian)
Xiamen, (Fujian)
Changchun (Jilin)
Karshi (Xinjiang)
Nanning (Guanxi)
Yilan, Nanning
Guiyang, Nanning
Additional observation sites: Jiaodong, Lhasa, Mount Wuzhi (Hainan)

Later in 1967, construction began of the network that would track the first Chinese Earth satellite, Dong Fang Hong. It was decided that there would be seven ground stations at strategic points on the Chinese land mass, the main control centre being in Xian,

Shaanxi (though in the event, Jiuquan fulfilled this function for the first launch). Each station would be equipped with tracking radars, tachymeters to receive the signals from the satellite, data processing systems, and control systems to send commands to the satellite. These relied largely on optical systems, but since then more sophisticated and accurate laser trackers have been introduced.

For the later series of communications satellites, China built more ground stations: in Nanjing (1975), Shijiazhuang (1976), Kunming (Sichuan) and Urumqi (1982), Beijing (1983, the central station) and Llasa, Tibet (1984).

China's main mission control is actually located in Weinan, 90 km north-east of Xian, in Shaanxi province, though it is normally referred to as Xian. At 34.3°N, 108.9°E, Xian is fed information from fixed tracking sites and up to three tracking ships at sea. Xian control was set up on 23 June 1967, though it was then called the Satellite Survey Department. Construction appears to have begun straight away, and the first computers were installed in 1968. Xian actually consists of two centres: a satellite telemetry and control centre, and an adjacent ground station. The control centre is located in 5,000 m high mountains, whereas the ground station is located 10 km distant on a high plateau. Between them are antenna farms and masts. In the autumn, as the communications dish swivels to track satellites in the sky, Chinese farmers may be seen gathering in wheat by hand on the terraces they share with the station. Mission controllers live nearby in modern apartment blocks. The mission control room comprises television screens, consoles, plotters and high-speed computers. The centre also includes a section devoted to following, calculating and predicting the orbital paths of all Chinese satellites in orbit at a given time. A core of 120 engineers runs the centre.

The ground station has a further 100 engineers and technicians, and has a key role in maintaining contact with geostationary satellites, with which it holds 30-minute communication sessions each day. In the 1980s, Zhou Enlai slogans were hung on the wall, reminding engineers to go about their tasks carefully and seriously.

A Landsat Earth resources station has been operating in Miyun (100 km north-east of Beijing) since 1986, with a processing centre in Beijing itself. It also receives data from the European Earth Resources Satellite (ERS) and the Japanese Earth Resources Satellite (JERS). Ground stations in Guangzhou and Urumqi receive data from the Feng Yun metsats and the American NOAA metsats.

In addition to controlling Chinese satellites in orbit, there is an important task in identifying and following satellites in orbit, be they Chinese or belonging to other countries but overflying China. Responsibility for tracking in China lies with the China Satellite Launch and Tracking General Control (CLTC), set up in 1966–67. CLTC comprises the Beijing Aerospace Command and Control Centre, the Satellite Maritime Tracking and Control (comships), the Institute for Special Equipment in Beijing, and the Institute for Tracking, Command and Control in Luoyang, down the Yellow River from Xian. CLTC monitors satellites in low Earth orbit and in 24-hour orbit.

COMSHIPS

The ground-based system is augmented by ship-based communications ships, or comships. They are important, because China does not have tracking stations in other

Yuan Wang comship.

countries. (The Soviet Union had an elaborate system of tracking ships, but the economic crisis forced Russia to sell them for scrap in the 1990s.) Comships were first used to track Chinese satellites from the sea in the 1970s, two oceanographic ships being used, the *Xiangyanghong 5* and *Xiangyanghong 11*.

The current tracking ships are called *Yuan Wang 1,2* and *3*. The first two *Yuan Wang* were ordered in September 1977 and delivered in 1978, being prepared for the tests of the Dong Feng 5 missile in May 1980. They are ocean-going in size, with 21,000 tonnes displacement, each being equipped with two 20 m wide communication dishes, and they carry out oceanic survey work when not participating in space tracking tasks. Their range is 21,000 km, and they can stay out at sea for 100 days at a time. In 1987 they were brought into dry dock for extensive refitting.

Yuan Wang 3 was commissioned in March 1995. This is a large ship of 20,000 tonnes displacement, 190 m long and with nine decks. Described as a scientific city at sea, it looks like a cruise liner. The *Yuan Wang* top deck is equipped with antennas, arrays and satellite dishes, with a helideck where weather balloons are launched. Below deck are computer and control rooms, much like mission control on land.

THE SPACE PROGRAMME AND THE CHINESE ECONOMY

There is no doubt that the development of a space programme forced the pace in a wide range of Chinese industrial technologies and products. It drove the development of computers, transistors, modern electronics, precision engineering, chemicals, welding, pumps and high-strength materials, to name just a few items. Evidence of the application of the space programme to the economy may be found in a variety of fields, such as important areas of material innovation, whether they came from the experience of handling low-temperature fuel or building high-temperature resistant heat shields: 220 new materials alone were used in the development of the Long March 3 third stage. In metallurgy, rocket development led to significant advances in high-temperature soldering, chemical milling of fuel and storage tanks, and the machining of titanium alloys. The welding skills used in rocket casings led to major advances in new forms of welding such as electron beam, plasma, laser and diffusion welding. Glass fibre reinforced plastics, used for the first time on the FSW recoverable cabin, had been an infant industry in China. The techniques used on developing the FSW were later applied in bottles and cylinders, tractors, locomotives, boats and household appliances.

The gains were not just in innovation, but also in standards. The requirements of quality control in rockets and satellites, in which parts were required to work flawlessly for years under extreme conditions, drove up standards in a range of industries. The tough quality requirements for rubber seals in rocket engines improved the standards of rubber seals used in more humble industrial and domestic situations.

By 1986, the Chinese were able to identify 1,800 industrial items of space spin-off, where techniques developed for the space industry had been used in the rest of the economy. By the same time, almost 5,000 awards had been made for space products and processes for standards, new products and inventions. These were some of the highlights:

- techniques developed for the automatic control of spacecraft at the Beijing Institute for Control Engineering were used to develop control systems for industrial film production;
- advance control technology was adapted directly to the production line system for the Beijing Packaging Company Cardboard Box Factory and for the Capital Machinery Factory's bread production line;
- telemetry control systems developed for the space industry were adapted for the navigational control system at Beijing airport and for automatic control of electric systems and reservoirs;
- satellite telemetry monitoring systems were adapted for telemetric monitoring of heart patients in hospitals and for an X-ray medical television system;
- temperature control systems devised for communication satellites have been used in cotton mills, swimming pools and the manufacture of ceramics, with considerable energy savings;
- vacuum test chambers, used to prepare satellites for the rigours of the space environment, have been used for the nuclear industry, and in advanced optics;
- cryogenic technology, mastered in connection with the Long March 3 upper stage, has been used in the development of refrigerators and night-vision optical equipment;

- television transmission systems, used on communication satellites, have been adapted for miniature television monitoring systems for the police, harbour, traffic control and airport authorities;
- the use of titanium as a material for rocket engines has been adapted to gas turbine engines; plasma spray coating developed for satellites is used as a permanent spray on electricity lines which previously had to be repainted every three years;
- precision machinery developed for rockets was used by the Chinese carpet industry to make new, automatic machines for weaving carpets with intricate designs;
- inertial guidance systems for rockets have been modified for installation on oil drills to measure their precise location under the ground;
- infrared satellite sensors have been fitted to trains to warn of bearings overheating and compromising safety.

INTERNATIONAL CONTACT

From 1956 to 1977, with the exception of the brief period of the Sino-Soviet accord, China developed its space programme relying almost entirely on its indigenous resources. During the period of rectification and reconstruction, Deng Xiaoping led a policy of openness and cooperation. International cooperation was subsequently promoted at a commercial level by the Great Wall Industry Corporation, at a scientific level by the Chinese Society for Astronautics, and at a ministerial level by the Ministry for the Space Industry.

A series of exchange visits and meetings kick-started the process in 1977–79, such activities taking place with Japan, the United States, and collectively and individually with the members of the European Space Agency. China's first international agreement was with France. The protocol agreed between the two countries covered cooperation in the areas of communications satellites, launchers, balloons and the surveying of natural resources. The Chinese were invited to watch the launch of an Ariane rocket. An agreement

Table 5.8. International cooperation agreements

Joint working groups	Cooperation agreements	Commercial arrangements to fly experiments	Joint programmes
United States	Germany	France	Brazil
Italy	Italy	Japan	Germany
	United States	Germany	
	France		
	Germany		
	Britain		
	Brazil		
	Russia		
	Ukraine		
	Europe		

with Italy shortly afterwards involved the use by the Chinese of an Italian communications satellite called Sirio, which was moved from its normal position in geostationary orbit (15°W) to 65°E, to test out ground stations in anticipation of China's first comsat. A memorandum was signed with the European Space Agency in 1986.

In the course of time, links were built up with over 40 countries. The standard procedure was for the first contacts to lead to bilateral visits, the exchange of minutes of meetings, followed by a protocol for cooperation initialled by the two governments. This led to structured contacts thereafter functioning at varying levels of intensity (permanent working groups, as with the United States, possibly being the most intensive). China has now entered cooperative arrangements with a number of countries, including collaborative programmes with Brazil (e.g. the Ziyuan project) and Germany (future solar observatory). These are detailed in Tables 5.8 and 5.9. In 1978 an exchange agreement with the United States was signed, and a joint working group with the Americans has been established since 1984. Two small Chinese student chemical and materials experiments flew on board the space shuttle (mission STS-42) in January 1992, and a further agreement with the United States in 1996 made provision for Chinese experiments to be flown on the space shuttle in 1998 and on the International Space Station in 2001.

In 1995, a $1.5 million agreement was made for France to supply an Earth terminal for the Russian–American COSPAS/SARSAT international search and rescue system in China. The most recent agreement was a cooperation agreement signed with Ukraine in 1997, Ukraine being the home of the biggest rocket factory in the world (NPO Yuzhnoye, Dnepropetrovsk), producer of the Zenith rocket and a range of satellites (such as Okean). The Harbin Polytechnic Institute bought a thermal simulator from the Ukrainians for $1 million. The agreement went ahead despite an incident just beforehand in which Chinese space experts were accused of spying in the Pivdenmach plant in Dnepropetrovsk.

Xi Chang launch centre has even received tourist visits. China has also welcomed people of Chinese extraction abroad, from the colonies of Hong Kong, Macao, what is euphemistically called 'the other side of the Taiwan straits' and the United States, including a shuttle astronaut of Chinese extraction, Taylor Wang. From 1985, China began to exhibit its commercial and scientific achievements in space at international exhibitions and big aerospace shows such as the biennial Paris air show.

Participation in the International Telecommunications Union, a specialised United Nations agency, was particularly important for China, since the body regulated international television and radio frequencies, including those of geostationary comsats. China did not obtain its seat there until 1972, at the expense of the Republic of China (Taiwan). It asked for and received three orbital positions for its geostationary comsats in 1977.

Table 5.9. China's membership of international space-related organisations

International Astronautical Federation
International Telecommunications Union
United Nations Committee on the Peaceful Uses of Outer Space
International Maritime Satellite Association (INMARSAT)
International Organization for Standardization

In 1980 China joined the United Nations committee on the exploration and peaceful uses of outer space. At the first meeting, China announced an allocation of $50,000 toward a conference on space science and technology in the Asia–Pacific region, which it would host (it was duly held in 1985). China has also signed the main international outer space treaties of the United Nations: those for the return of stranded astronauts, responsibility for damage caused by space objects and the registration of objects launched into space. Finally, the Beijing Marine Communications and Navigation company is the Chinese member of the International Maritime Satellite Association (INMARSAT). It sells and leases INMARSAT lines and equipment to foreign and domestic customers in China.

ASSESSMENT AND CONCLUSIONS

The Chinese space industry is an important part of the Chinese economy, worth about £850 million ($1.35 billion) a year. In financial terms, China is the fifth largest spending nation in the world space league. The numbers employed are considered to be in the order of 200,000 people, half of them skilled, and a tenth of these being graduates. The benefits of the Chinese space programme have been applied throughout the country's economy. Space technology has led to the development of higher standards and new processes in a wide range of industry. Whilst satellites and launchers are the visible part of the space programme, a considerable infrastructure is necessary to maintain such an industry. This infrastructure includes four rocket launching bases, several large industrial companies, several hundred research institutes, and a variety of sophisticated test and research centres. Launch sites and launchers represent one of the biggest areas of investment in the programme, and they are the theme of the next chapter.

6

Launch sites, launchers and engines

LAUNCH SITES

Perhaps the most important and expensive infrastructural element of any space programme is its launch site facilities, of which China has four. The first, Jiuquan, was built in northern China for China's first satellite, Dong Fang Hong. The second, Xi Chang, was built in south-western China for launches to equatorial orbit. The third, Taiyuan, was built for launches to polar orbit. Finally, there is a minor launch site for sounding rockets (Haikou)[32]. For the sake of completeness, one should mention the military-only launch site in Harbin, Manchuria, used for the Dong Feng 4 and 5, and presumably for its replacements, the DF-31 and DF-41. There have been occasional discussions about the construction of equatorial Long March launch sites abroad, though these have never come close to project design. These discussions centred on a collaborative project with Brazil for an equatorial launch site, and with Indonesia on Biak island.

Table 6.1. Launch sites

Name	Location	First flight	Launchers
Jiuquan	Shuang Cheng-tzu, 42.7°N, 99.97°E	24 Apr 1970	CZ-1, 2, 2C, 2D, 1D, FB-1
Xi Chang	Sichuan, 28.25°N,102.02°E	29 Jan 1984	CZ-3, 2E, 3A, 3B
Taiyuan	Shaanxi, 38.8°N, 111.5°E	6 Sep 1988	CZ-4A, 2C-SD, 4B
Haikou	Hainan, 19.31°N, 109.8°E	19 Dec 1988	Weaver Girl sounding rocket

Although Chinese launch centres rely on rail, as with Russia, in practice rocket stages are often brought the final stages to the pad by road. In America, rocket stages are assembled vertically indoors, but in China this is done outside. Of the three main launch sites, the busiest is Jiuquan. Table 6.2 lists the total number of launch attempts (including failures) from each.

Table 6.2. Chinese launch attempts

Jiuquan	26
Xi Chang	25
Taiyuan	4
	55

Jiuquan

Jiuquan launch site is located in the Gobi desert in north-west China, in an environment similar to Baikonour cosmodrome in Kazakhstan. The cosmodrome covers an area of 5,000 km^2. Storms whip up the desert sands from time to time, and dunes often creep towards the railway line. The average annual rainfall is 700 mm and the thin soil is a light dusty brown shade. Jiuquan is on the river Ruoshui, which runs northwards and is very seasonal. Two thousand years ago, Jiuquan was a battleground where the Han dynasty erected fortifications to keep out the Huns. Subsequently it became part of the north-western silk road – a long trading route which, thousands of kilometres further west, passed along another cosmodrome in the Kazakhstan steppe. At Jiuquan, there are few bushes, only brown camelthorn, and few wild animals, mainly yellow goats and wild deer. Not far from the launch site, Mongolian herdsmen may be seen from time to time minding their sheep.

Construction of Jiuquan began in April 1958 for the launch of the R-2 Soviet rocket which eventually took place in November 1960. In advance of the launch of China's first satellite, Dong Fang Hong, there was a substantial expansion of the facility.

Western experts became familiar with Jiuquan in 1976 when the first Landsat Earth resources satellite photographs became available (though US military intelligence would have had its own pictures long before). American aerospace experts visited the site in 1979. There are two launch areas: area no. 3 (the first, used for military missiles, with one pad) and area no. 2, the second one built, used for the space programme, consisting of two pads. Missiles are fired from a crude cement pad on area 3 to Anxi in Xinjiang (short-range), to Korla, Taklimakan (medium-range) and Minfeng and Khotan in Yecheng (for long-range missiles).

On the west side of the river, in close proximity to each other, are the two pads of area 2,416 m apart, for the Long March 1, 2 and Feng-Bao launchers, now the primary launch site area. Both pads are served by one 55 m tall, 1,400 tonne gantry moving on 17 m wide rails between the two. One pad is called #5020 (designed for the Long March 1) and the other #138 (designed for the Feng Bao, DF-5 and Long March 2). Each has a flame trench. #5020 has a 37 m umbilical tower which provides fuel, gas and electricity right up to the final moments of the countdown. Pad #138 has a 40 m umbilical tower with 11 floors, a rotary launching stand with four supporting arms for holding the rocket and a 19 m flame trench. The gantry has a clean room for final payload installation. #138 can handle hot tethered test firings, in which the rocket fires all its engines but does not take off.

A domed launch control blockhouse is 200 m distant, 10 m underground, from which controllers watch the launch by TV and periscope. Fuels are stored in underground

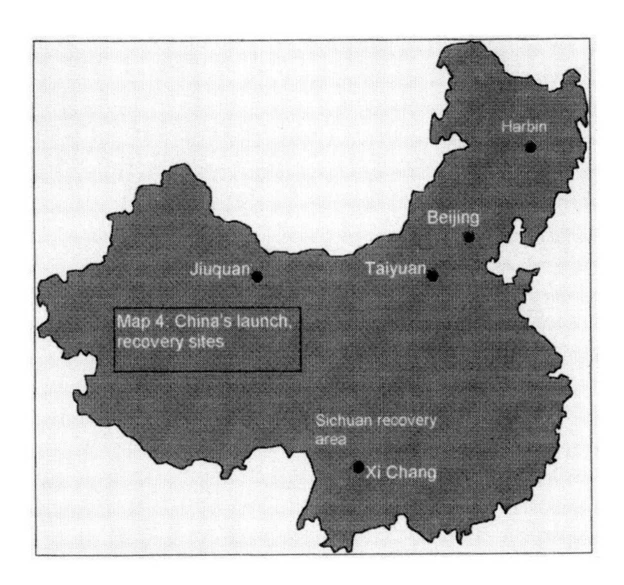

Map of China's launch and recovery sites.

bunkers. Due south lies the Huxi Xincun range control centre, two assembly buildings, blockhouses and electricity station.

The assembly buildings are 140 m long, and adjacent to these are 25 test rooms, including a clean room. There are barracks for the militia who assist in the launchings (Russian rockets are also launched by troops) and four-storey Soviet-style flats for other workers involved in the launchings. Willow and white poplar trees are planted around the buildings and walkways to provide wind breaks and colour. The main facilities are connected by railway lines and tarmacadamed roads. The general headquarters lie further south along the river, as does the airport and, still further south, high mountains. Launches from Jiuquan curve over to the south-east and take orbital inclinations of 56°–70°, otherwise they will fly over Mongolia to the west or Russia to the north.

Jiuquan can handle two launch campaigns simultaneously, and has been identified as the site for China's first manned spaceflight. If this is the case, then it will more than likely require a new launch pad able to take the more powerful Long March 2E. However, there is no confirmation that construction of a LM-2E pad has yet started. In the early days, conditions in Jiuquan must have been very primitive. Now, the centre has a fully equipped railway station, hospital and library, and 20,000 people live and work there.

Xi Chang

When seeking a launch site closer to the equator, Xi Chang launch centre was first identified in 1972 after a number of possibilities had been considered. Altogether, 80 sites were surveyed, shortlisted to 16 before Xi Chang was selected. The final decision appears to have been made by Zhou Enlai personally. Xi Chang was constructed in the course of 1978–1982. The first launch rehearsal was conducted in 1983, with the centre opening in January 1984.

Xi Chang is located 1,800 m high in mist-shrouded mountains, near Mount Lijang in the direction of Burma. It must be one of the most scenic launch sites in the world. Nearby are rice paddies and grazing buffalo, to the north lie mountains, and to the south lie lakes. The launch site is 65 km from Xi Chang city and 270 km from the city of Chengdu by rail and road. Xi Chang is on the south-western silk road which started at Chengdu and headed through Xi Chang into Burma (250 km distant) and India. It is a main passing point for migratory birds.

General view of Xi Chang launch site.

Western commentators have expressed surprise at the selection of a launch site so far inland, in difficult terrain, with poor communications facilities in a relatively populated rural area. The Chinese subsequently explained that an inland site was preferred, because coastal sites, though more southerly, were vulnerable to attack by China's enemies.

Xi Chang city has a combined civil and military airport – one now capable of taking jumbo jets – and is the headquarters of the launch site organisations. Xi Chang is reached by following the Kunming–Chengdu railway northward along the valley floor of the Anning river until a single branch line turns west into the launch site valley, which is entered by passing a Soviet-style people's memorial. Visitors arriving there drive along roads cluttered by bicycles, water buffalo and farm workers carrying chickens and vegetables to market. After passing a communication centre, technical centre and command and control centre, to the right of the railway is the first launch pad, for the Long March 3, served by a large 900 tonne, 77 m tall gantry which has 11 work levels and a crane. A cement flame trench has been constructed to take away the flames of the rocket on take-off.

To the left of the railway, 100 m away, is a second launch pad, linked by a mobile service tower, constructed subsequently for the Long March 2E, 3A, 3B and 3C. It was built in the course of 14 months, and was first used for the CZ-2E in 1990. This second pad has a huge 4,580 tonne service tower, 97 m tall, with 17 work levels. Just 80 minutes before launch, the tower moves back to a distance of 130 m. Chinese satellites were originally first stacked on the pad, the shroud then being clamped on top. This exposed the satellite to dust and humidity, but Xi Chang now has an air-conditioned clean room on the upper gantry work level.

Also to the left of the railway line lie buildings for storing launchers, the various stages and payloads, though they are finally assembled vertically on the pad by crane. The launch towers are protected by 100 m high lightning rods. Around the gantries are fuelling lines – one set to keep the liquid hydrogen third stage topped up, a second to provide helium which pressurises the fuel tanks, and a third for storable fuels. Liquid hydrogen is topped up in the third stage until just three minutes before lift-off.

Launches out of Xi Chang take a curving trajectory to the south-east, flying over southern Taiwan, north over the Philippines and towards the equator. The ascent is tracked from either side by ground stations in Yibin and Guiyang, Nanning. Satellites are still over China when they reach space. The countdown is carried out in a blockhouse close to the pad, but the overall operation and the subsequent flight are monitored from the launch control centre 6,000 m from the launch pad in a deep gully[33]. The launch control centre comprises a large gymnasium-size room with walls of consoles and a large, 4 m × 5.3 m visual display at the front. The centre is not hardened against explosions or falling debris. There is an observation room which can accommodate 500 people at a time. Laser theodolites, set in domes, track rockets as they ascend to orbit.

A launch campaign in Xi Chang takes 40 days. The rocket is first delivered by rail into a transit hall measuring 30.5 m × 14 m, before being brought into a much larger assembly room of 91.5 m × 27.5 m. Payloads are checked out in a clean room measuring 42 m × 18 m, called the non-hazardous operations building, where temperatures and humidity are kept within tight limits. The stages and payloads are then transferred to the hazardous operations and fuelling building where the fuelled solid-rocket stages are installed. Final checks take place in a final checkout and preparation building. The site also has an X-ray facility to check any equipment against cracks. The rocket stages are trolleyed to the pad, one by one, before being assembled vertically.

Taiyuan

Taiyuan launch centre was developed as a launch site for the Dong Feng missile. The base is set in gently rolling hills 90 km south-west of Beijing. Construction was authorised by Mao Zedong, Liu Shaoqi and Zhou Enlai in March 1966. The early construction crews had a difficult time, surviving on Chinese cabbage and potato, but they succeeded in their main task of building a 76.9 m high launching platform of 11 stories. The first Dong Feng was launched from there on 8 December 1968. Taiyuan, built on loamy yellow rocks out-

Taiyuan launch site.

Long March 4 night launch from Taiyuan.

side the city, was used for missile tests in the 1970s (for the submarine-launched ballistic missile) but was not brought into the space programme until the early 1980s. The first static rocket tests were carried out in September 1984.

Taiyuan was eventually introduced as a national launch site in 1988, with the first launch of the Feng Yun weather satellite on the Long March 4. It has a single pad for rocket launches. Initially, only the Long March 4 was used here, but the site is to support the Long March 2C-SD launches of Iridium comsats, the first of which took place in September 1997. The first photographs of the site were not released until 1990. Taiyuan is much less well known in the west than Jiuquan or Xi Chang, though with Iridium launches it will become more familiar.

Haikou

Construction of Haikou, Hainan, launch site began in 1986, and it was commissioned in 1988. It was used for the launch of the four Weaver Girl 1 sounding rockets in 1988 and the single Weaver Girl 3 launch in 1991. The site is intended to serve as a launching platform for sounding rockets able to reach between 120 km and 300 km. The site consists of a launch pad, underground control centre, tracking and payload retrieval systems.

LAUNCHERS

China has developed two families of launchers: the Long March, known as the Chang Zheng, or CZ, and the Feng Bao (FB, or Storm). The Long March family is divided into four series – 1, 2, 3 and 4. The Feng Bao launcher was used from 1971 to 1981, and is no longer in service. A considerable amount is now known in the West about Chinese launchers, both because they are commercially promoted in the West and due to the work of analysts such as Clark[34].

When first encountered, the numbering pattern for Chinese launchers can be quite confusing. Although the Long March 3 series (the 3, 3A and 3B) is designed for geostationary orbit, one of the 2 series (2E) is also used for these missions. Thankfully, the Chinese are visually helpful in enabling us to identify rocket launchers: their white-painted rockets invariably have the launcher type painted in big red letters in English script on the side.

Although, to an outsider, all rockets, being rocket-shaped, appear to have the same means of propulsion, in fact there are many important distinctions between them. Firstly, rockets may use either solid fuel or liquid fuel. Solid-fuel rockets operate on the same principle as fireworks. A grey sludge-like chemical is poured into a rocket container, and when the nozzle is fired the stage burns to exhaustion. Solid rockets are very powerful, but their main disadvantage is that they cannot be turned off: they simply burn out. They are less precise and less safe.

Liquid-fuel rockets are more complex. They have two tanks – a fuel tank and an oxidizer. Both are pressurised, and fuel is injected, at great pressure, into a rocket engine, which is ignited. With liquid-fuel engines, the level of thrust may be varied and the engine may be turned off and restarted. This system is more complex but more versatile and, from a manned spaceflight perspective, safer. Liquid-fuel rockets may be divided into three sub-categories, according to the type of fuel used. Most Russian and American civil rockets have used kerosene (a form of paraffin) or its variants as a fuel. These are powerful fuels, but they degrade if they are kept in a rocket for more than a few hours at a time. If a launching is missed, the fuels have to be drained and reloaded – a tedious and time-consuming process. From the 1960s, Russian and American military rockets began to use storable propellants, generally based around nitric acid and UDMH (unsymmetrical dimethyl hydrazine). The advantage of storable propellants is that they can be kept in rockets for long periods before they are fired – a necessity when military rockets must be kept in a constant state of readiness. Their disadvantage is that such fuels are highly toxic, presenting hazards for launch crews and horrific consequences in an explosion[35]. Finally, there is liquid hydrogen, which is enormously powerful but has to be kept at extremely low temperatures.

Most modern Chinese rockets use storable propellants for their main stages, and solid rocket boosters for their small upper stages. The Chinese introduced a hydrogen-fuelled upper stage with the Long March 3 in 1984. The main centres for the development of Chinese rockets are the Chinese Academy of Launcher Technology (CALT) in Beijing, the Shanghai Academy of Space Technology (SAST), and the Liquid Rocket Engine Co. in Shaanxi.

Table 6.3. Chinese launchers and their design bureaux

Launcher	1st stage	2nd stage	3rd stage
Long March 1	CALT	CALT	Solid Engine Research Institute
Feng Bao 1	SAST	SAST	
Long March 2C	CALT	CALT	
Long March 2E, 2D	SAST	SAST	
Long March 3, 3A, 3B, 3C	SAST	SAST	CALT
Long March 4	SAST	SAST	SAST

Adapted from Clark, *Chinese launch vehicles*, 8

The rocket designers

Long March 1	Ren Xinmin
Long March 2	Tu Shoue
Long March 3	Xie Guangxuan
Long March 4	Sun Jinliang
Feng Bao 1	Shi Jinmiao

Long March 1 (CZ-1): launcher of China's first two satellites

The Long March 1 launched China's first two satellites into orbit: Dong Fang Hong in 1970 and Shi Jian 1 the following year. The Long March 1 is essentially a three-stage version of the Dong Feng 4 medium-range missile developed over 1965–70, a weapon

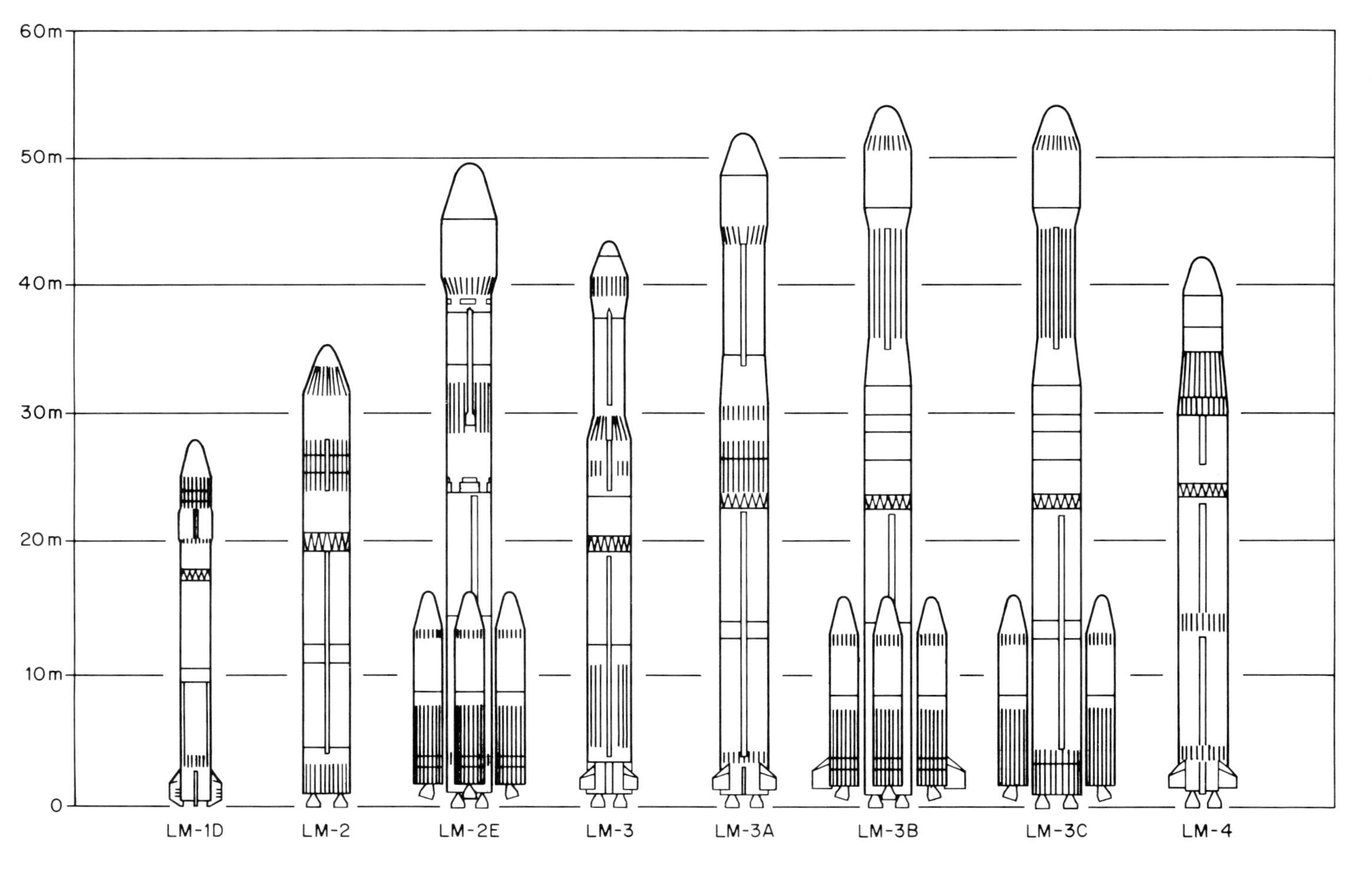

The Long March family of launch vehicles.

planned to hit targets as far away as the mid-Pacific. On top of the DF-4, a small third stage solid-rocket motor stage was fitted to put the two satellites into orbit.

Table 6.4. Long March 1 launcher (CZ-1)

Dimensions	29.45 m long, 2.25 m diameter
Weight at lift-off	81.6 tonnes
Capability	300 kg to 440 km, inclination 70°

Flight record

24 Apr 1970	Dong Fang Hong 1	173 kg	China's first satellite.
3 Mar 1971	Shi Jian	221 kg	Scientific satellite.

	1st stage	2nd stage	3rd stage
Dimensions	17.835 m × 2.25 m	5.35 m × 2.25 m	4 m × 0.77 m
Weight	64.1 tonnes	14.85 tonnes	N/A
Engines	Four YF-1A	YF-3	GF-02
Fuels	UDMH, nitric acid	UDMH, nitrogen tetroxide	Solid
Thrust	104 tonnes	30 tonnes	3 tonnes
Burn time	140 s	120 s	N/A

Three subsequent versions of the Long March 1 have been proposed. The CZ-1C and the CZ-1M were proposed in the mid-1980s as improved versions of the Long March 1, each with more powerful third stages, but neither was developed nor flown. A smaller version, the CZ-1D, is now on offer commercially. The Long March 1 series still has an honoured place in China's astronautical history: launcher of China's first two satellites, with a 100 per cent reliability record.

Long March 2 (CZ-2): launcher of China's recoverable satellites

The Long March 2 series must be subdivided into four types:

Long March 2A, which made one failed launch in 1974 and was then improved;
Long March 2C, which introduced the recoverable satellite programme in 1975;
Long March 2D, which introduced heavier recoverable satellites from 1992; and
Long March 2E, a radically more powerful version, used for missions to 24-hour orbit.

(The Long March 2B was a cancelled design for a version to carry a small payload to 24-hour orbit). The Long March 2A made one flight, in November 1974, which ended disastrously. This was the first attempt to place a recoverable satellite into orbit. The series of extensive improvements to the Long March design which followed saw the new rocket being named the Long March 2C. This is one of the most successful of China's rockets. From 1975 to 1993, it flew 14 times, each launch being successful. One satellite failed to return, but this was not the fault of the launcher.

Table 6.5. Long March 2A, 2C launcher (CZ-2A, 2C)

Dimensions	32.57 m long, 3.35 m diameter
Weight at lift-off	192.15 tonnes
Capability	2,500 kg to 170–300 km, 56° to 67°

Flight record

Date	Payload	Mass	Notes
5 Nov 1974	FSW	1,800 kg	Failed after 20 seconds; vehicle destroyed
26 Nov 1975	FSW 0-1	1,790 kg	First recovery by China of payload from orbit.
7 Dec 1976	FSW 0-2	1,790 kg	
26 Jan 1978	FSW 0-3	1,810 kg	
9 Sep 1982	FSW 0-4	1,780 kg	
19 Aug 1983	FSW 0-5	1,840 kg	
12 Sep 1984	FSW 0-6	1,810 kg	
21 Oct 1985	FSW 0-7	1,810 kg	
6 Oct 1986	FSW 0-8	1,770 kg	
5 Aug 1987	FSW 0-9	1,810 kg	
9 Sep 1987	FSW 1-1	2,070 kg	
5 Aug 1988	FSW 1-2	2,130 kg	
5 Oct 1990	FSW 1-3	2,080 kg	
6 Oct 1992	FSW 1-4	2,060 kg	Carried Swedish Freja satellite.
8 Oct 1993	FSW 1-5	2,100 kg	Recovery failed; crashed into Atlantic, 1996.
1 Sep 1997	Iridium	650 kg × 2	Test of CZ-2C-SD for Iridium comsats.
8 Dec 1997	Iridium	650 kg × 2	Iridium comsats.

	1st stage	2nd stage
Dimensions	20.52m × 3.35m	7.50m × 3.35m
Weight	151.55 tonnes	38.5 tonnes
Engines	Four YF-20	YF-22
Fuels	UDMH, nitrogen tetroxide	UDMH, nitrogen tetroxide
Thrust	284 tonnes	73.2 tonnes
Burn time	130 s	130 s

All flights to date have been made from Jiuquan. The Long March 2C series might have ended in 1993, had it not been for the American Motorola company, which booked the Long March 2C for 11 launches of its Iridium global telecommunications satellite

(22 satellites in total). The 2C is adapted with a longer second stage (2 m longer) and what is called a 'smart dispenser' (SD), designed to spring the small comsats into orbit. The Taiyuan site is used for these flights, destined to fly into a new orbit, 700 km altitude at inclination 58°. This launcher is referred to as Long March 2C-SD. A test of the SD was made on 1 September 1997, and it appears to have been entirely successful. The first full mission was flown on 8 December.

Long March 2D (CZ-2D)

The Long March 2D was introduced in 1992 to carry the heavier, third generation of FSW recoverable spacecraft. The payload of the Long March 2D is 3,400 kg to low Earth orbit, about 400 kg more than the Long March 2C. The launcher is slightly heavier, with improved performance in a number of areas. Three launches have been made, all successful and all from Jiuquan.

Table 6.6. Long March 2D launcher (CZ-2D)

Dimensions	37.72 m long, 3.35 m diameter
Weight at lift-off	228.4 tonnes
Capability	3,400 kg to 200 km

Flight record

Date	Payload	Mass	Result
9 Aug 1992	FSW 2-1	2,590 kg	Recovered after 16 days
3 Jul 1994	FSW 2-2	2,760 kg	Recovered after 13 days
20 Oct 1996	FSW 2-3	2,970 kg	Recovered after 15 days

	1st stage	2nd stage
Dimensions	24.92 m × 3.35 m	7.92 m × 3.35 m
Weight	187.7 tonnes	36.2 tonnes
Engines	Four YF-20B	YF-22B
Fuels	UDMH, nitrogen tetroxide	UDMH, nitrogen tetroxide
Thrust	302 tonnes	N/A
Burn time	154 s	94.4 s

Long March 2E: China's heaviest launcher

Although the Long March 3 launcher is primarily associated with the missions to geostationary orbit which began in 1984, a version of the Long March 2 was adapted for flights to 24-hour orbit. This is the Long March 2E (CZ-2E). Approval of the CZ-2E project was made at a State Council conference chaired by premier Li Peng in 1988. Production took 18 months, the launcher making its debut in July 1990. It is China's heaviest launcher, weighing 463 tonnes on the pad (though it is not its most powerful; that distinction belongs to the 3B version).

Strictly speaking, the Long March 2E is not a geostationary launcher. It has two stages which put 9 tonnes into low Earth orbit. This payload can be used to carry a powerful solid-rocket booster (a perigee kick motor) to send a communications satellite on its way to geostationary orbit. Once there, the satellite's own motor can adjust the orbit to make a perfect 24-hour orbit for communications (using an apogee kick motor). Equally, it has not escaped the notice of Western observers that 9 tonnes is an adequate size for a manned spacecraft (the Russian Soyuz is in the order of 7 tonnes).

To adapt the original Long March 2 for flights to geostationary orbit, four liquid-fuel strap-on boosters were fitted. Their use is not unusual in rocketry, as they have been used on the famous American launcher, the Delta, and Europe's Ariane. What is unusual is that China uses *liquid* fuel strap-ons, unlike other countries, which use solid-rocket motors to

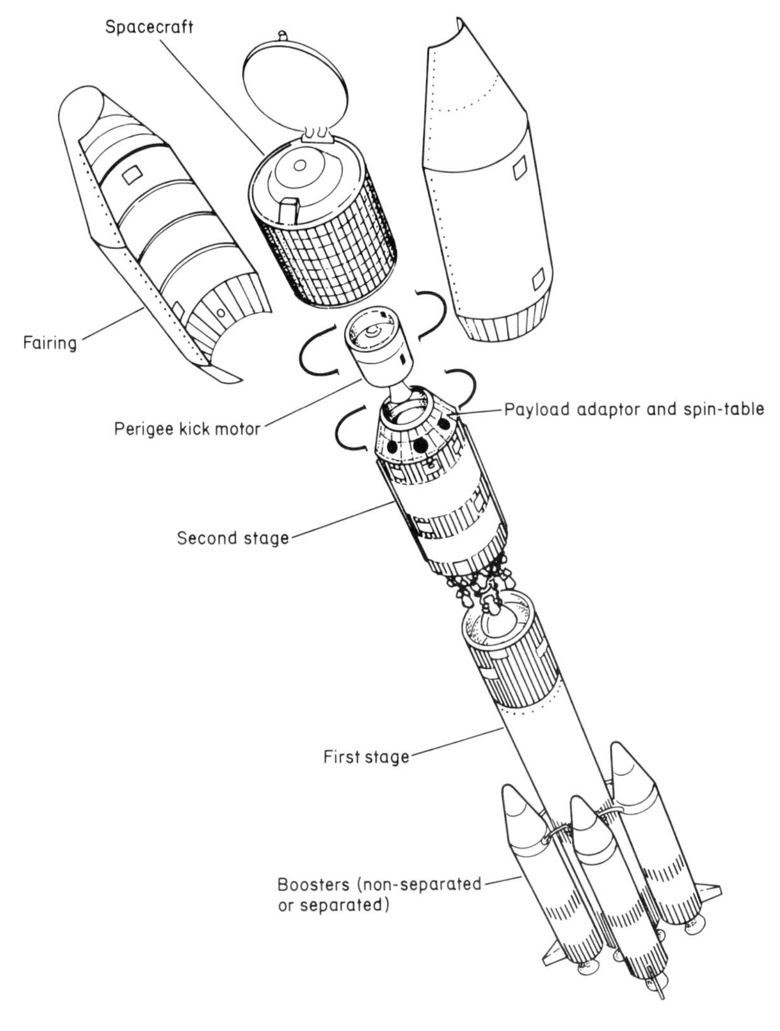

CZ-2E schematic view.

provide additional thrust. For the CZ-2E, the Chinese added four strap-on boosters with YF-20B engines. At lift-off, eight engines light up simultaneously – the four core engines of the first stage and the four strap-ons. The noise is tremendous, reaching 142 decibels.

In a typical launch profile, the Long March 2E drops its four strap-on rockets when they burn out at 2 min 5 s. Staging takes place at 2 min 40 s, the second, top stage entering orbit at 10 min. At this point, the solid-rocket motor is expected to fire the payload into geostationary transfer orbit. Two such motors have been used: the Chinese-built perigee kick motor – the EPKM – and the American Star 63 on American satellite payloads.

The CZ-2E was first launched in July 1990, when it lifted the Pakistan satellite Badr into orbit. However, the attempt to fire the EPKM to put a model comsat into geostationary transfer orbit failed, the EPKM firing the wrong way and bringing the model out of orbit. The subsequent use of the EPKM on Asiasat and Echostar was entirely successful. The use of the American Star 63F kick motor has been controversial. Although the Star 63F performed as advertised on its first and third mission (the Australian Optus B-1 and B-3), the second and fourth missions (Optus B-2 and Apstar 2) were lost.

Table 6.7. Long March 2E launcher

Dimensions	51.2 m long, 3.35 m diameter
Weight at lift-off	463.7 tonnes
Capability	9 tonnes to low Earth orbit; 3.4 tonnes to geostationary transfer orbit

Flight record

16 Jul 1990	Badr, Aussat model	7,405 kg	Badr entered orbit; model failed.
13 Aug 1992	Optus B-1	3,000 kg	Australian comsat.
21 Dec 1992	Optus B-2	3,000 kg	Only debris entered orbit.
27 Aug 1994	Optus B-3	3,000 kg	Australian comsat.
25 Jan 1995	Apstar 2	3,000 kg	Exploded after 70 s.
25 Nov 1995	Asiasat 2	3,379 kg	Hong Kong comsat.
28 Dec 1995	Echostar 1	3,288 kg	American comsat.

	First stage	Strap-ons	Second stage	Third stage
Dimensions	23.7 m long, 3.35 m diameter	16.017 m long, 2.25 m diameter	15.52 3m long, 3.35 m diameter	3.07 m long, 1.7 m diameter
Weight	195.7 tonnes	41 tonnes (× 4)	93.5 tonnes	5.9 tonnes
Engines	Four YF-20B	Four YF-20B	YF-22B	SPTM-17
Fuels	UDMH, nitrogen tetroxide	UDMH, nitrogen tetroxide	UDMH, nitrogen tetroxide	Solid
Thrust	300 tonnes	300 tonnes	93.5 tonnes	21 tonnes
Burn time	160 s	125 s	300 s	75 s

Long March 3

The Long March 3 was introduced in order to give China the capability to fly to geostationary orbit. It was introduced in January 1984, although the satellite was left stranded in low Earth orbit. Since then, the rocket has been used for both domestic and foreign communications satellite launches. In June 1997, the Long March 3 successfully launched the first geostationary meteorological satellite, Feng Yun 2. The H-8 upper stage is broadly comparable to the American Centaur, but more powerful than Europe's Ariane. The first two stages of the Long March 3 are manufactured in the Xin Zhong Hua factory in Shanghai. About 1,000 people, including 400 engineers, work on rocket production. The centre can handle up to six Long Marches at a time.

The Long March 3 has been launched 12 times. It has clearly experienced problems with its third, hydrogen-fuelled stage, for it has failed on three occasions, even though the

Asiasat 1 launch on 7 April 1990.

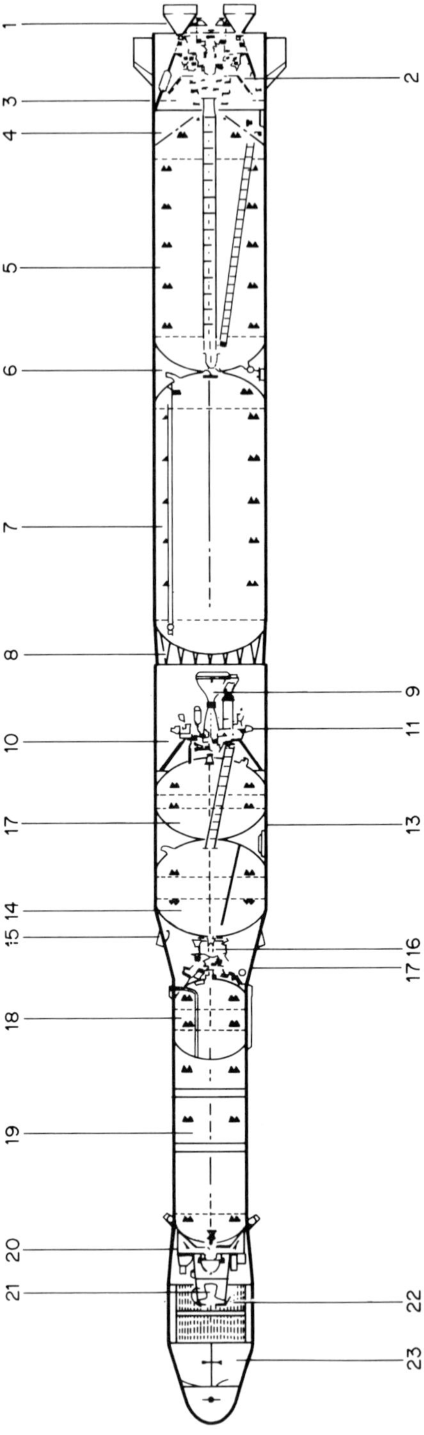

Cutaway drawing of the CZ-3 launch vehicle. Key to labels: 1 - first substage engine, 2 - first substage servo-mechanism, 3 - first substage tail section, 4 - aft skirt, 5 - fuel tank, 6 - inter-tank section, 7 - oxidiser tank, 8 - interstage truss, 9 - second substage engine, 10 - interstage section, 11 - second substage servo-mechanism, 12 - fuel tank, 13 - inter-tank section, 14 - oxidiser tank, 15 - retrorocket, 16 - third substage engine, 17 - interstage section, 18 - liquid oxygen tank, 19 - liquid hydrogen tank, 20 - vehicle equipment bay, 21 - payload/vehicle adapter, 22 - payload fairing (model A).

precise causes for the failures seem to have varied. All these failures occurred on domestic comsat missions. The Chinese will undoubtedly hope that with the return to flight of the Long March 3 in June 1997, they can put these problems behind them.

Table 6.8. Long March 3 launcher (CZ-3)

Dimensions	43.85 m long, 3.35 m diameter
Weight at lift-off	201.7 tonnes
Capability	1,400 kg to geostationary orbit

Flight record

Date	Payload	Mass	Notes
29 Jan 1984	Shiyan Weixing	1,000 kg	Third stage failed to ignite.
8 Apr 1984	Shiyan Tongbu Tongxin Weixing	1,000 kg	First successful mission to 24-hour orbit.
1 Feb 1986	Shiyong Tongbu Tongxin Weixing 1	1,000 kg	First operational 24-hour satellite.
7 Mar 1988	Shiyong Tongbu Tongxin Weixing 2	1,025 kg	
22 Dec 1988	Shiyong Tongbu Tongxin Weixing 3	1,025 kg	
4 Feb 1990	Shiyong Tongbu Tongxin Weixing 4	1,024 kg	
7 Apr 1990	Asiasat 1	1,422 kg	Commercial launch (Hong Kong); formerly Westar 6 satellite.
28 Dec 1991	Shiyong Tongbu Tongxin Weixing 5	1,025 kg	Third stage failed to ignite.
21 Jul 1994	Apstar 1	1,383 kg	Commercial launch (Hong Kong).
4 Jul 1996	Apstar 1A	1,385 kg	Commercial launch (Hong Kong).
18 Aug 1996	Zhongxing 7	1,200 kg	Third stage failed to ignite.
10 Jun 1997	Feng Yun 2	1,380 kg	First geostationary metsat.

	1st stage	2nd stage	3rd stage
Dimensions	20.219 m × 3.35 m	79.707 m × 3.35 m	7.484 m long × 2.25 m
Weight	149.45 tonnes	39.5 tonnes	10.5 tonnes
Engines	Four YF-20	YF-22	YF-73
Fuels	UDMH, nitrogen tetroxide	UDMH, nitrogen tetroxide	Liquid hydrogen, liquid oxygen
Thrust	284 tonnes	73.2 tonnes	4.5 tonnes
Burn time	130 s	130 s	13 min 20 s

Long March 3A taking off.

Long March 3A: double the performance

The Long March 3A was the first variant of Long March 3, being introduced 10 years later. Compared with the Long March 3, it offered substantially improved performance and is able to place about twice the weight in geostationary. It has a stretched first stage and bigger third stage (called the H 18). This launcher has been used three times, successfully on each occasion. The third stage is entirely redesigned and carries two YF-75 engines (rather than the two on Long March 3). Ten small engines were fitted to the third stage in an attempt to settle the propellant before its second ignition. The Long March 3A has a new, advanced digital computer system.

The third-stage hydrogen engine was developed by Zeng Guang Shang, Wang Heng and Shen Weigou. It took them five years, working in an open assembly plant which let in cold north-west winter winds and the scorching sun of summer. The welding of the engine chamber produced particular difficulty, with designers breaking into tears in their frustration. A team of seven women designers, led by Fang Rongchu, was assigned to the electronic control system of Long March 3A, the system passing its test in a 480-hour soak in vacuum and thermal chambers.

Table 6.9. Long March 3A launcher (CZ-3A)

Dimensions		52.32 m long, 3.35m diameter	
Weight at lift-off		238.5 tonnes	
Capability		2,300 kg to geostationary orbit	
Flight record			
8 Feb 1994	Shi Jian, KF-1	2,000 kg	Science mission
29 Nov 1994	Zhongxing 6	2,230 kg	Domestic comsat; failed
8 May 1997	Dong Fang Hong 3-2	2,000 kg	Domestic comsat

	1st stage	2nd stage	3rd stage
Dimensions	23.075 m × 3.35 m	11.256 m × 3.35 m	7.484 m long × 3 m
Weight	178.5 tonnes	35.1 tonnes	21.1 tonnes
Engines	Four YF-20B	YF-22B	YF-75
Fuels	UDMH, nitrogen tetroxide	UDMH, nitrogen tetroxide	Liquid hydrogen, liquid oxygen
Thrust	302 tonnes	73.2 tonnes	8 tonnes
Burn time	146 s	110 s	8 min

Long March 3B: China's most powerful launcher

The Long March 3B is the most powerful rocket in the Chinese armoury, equivalent to Russia's Proton and Europe's top version of the Ariane 4. In essence, it adds the four strap-on rockets from the CZ-2E to the CZ-3A to get a payload of 4.8 tonnes to geostationary orbit.

This weight is comparable with the best rockets operated by Russia, Europe and the United States. It has larger propellant tanks and better computer systems than its predecessors. Although not as heavy as the Long March 2E (425 tonnes compared with 463 tonnes), it is more powerful, generating 5,923 kN of thrust (compared with 5,919 kN for the CZ-2E).

The Long March 3B has made only two flights. The purpose of its maiden flight was to carry into orbit the Intelsat 708, but it crashed only 2 s into its mission. No test flight had been carried out, largely because the CZ-3B had so many features in common with previous Long March launchers that it was not considered necessary. This judgement may have been a mistake. Repairing the problems took 18 months, the first successful mission being the launch into 24-hour orbit of the Philippine satellite Agila in August 1997. Table 6.11 lists the capacities of the various versions of Long March to geostationary orbit.

Table 6.10. Long March 3B launcher (CZ-3B)

Dimensions	52.32 m long, 3.35 m diameter
Weight at lift-off	425 tonnes
Capability	2,300 kg to geostationary orbit

Flight record

14 Feb 1996	Intelsat 708	4,576 kg	Crashed 2 s after launch, with fatalities.
20 Aug 1997	Agila	3,770 kg	24-hour orbit.
16 Oct 1997	Apstar 2R	3,747 kg	24-hour orbit.

	First stage	Strap-ons	Second stage	Third stage
Dimensions	23.075 m long, 3.35 m diameter	16.017 m long, 2.25 m diameter	13.75 m long, 3.35 m diameter	3.07 m long, 1.7 m diameter
Weight	180.3 tonnes	41.2 tonnes (× 4)	55.6 tonnes	21.7 tonnes
Engines	Four YF-20B	Four YF-20B	YF-22B	YF-75
Fuels	UDMH, nitrogen tetroxide	UDMH, nitrogen tetroxide	UDMH, nitrogen tetroxide	Solid
Thrust	302 tonnes	305 tonnes	73.2 tonnes	16 tonnes
Burn time	146 s	125 s	185 s	75 s

Table 6.11. Capacities of Long March series to geostationary orbit

Launcher	tonnes
Long March 2E	3.4
Long March 3	1.4
Long March 3A	2.3
Long March 3B	4.8
(Long March 3C	3.5 (projected))

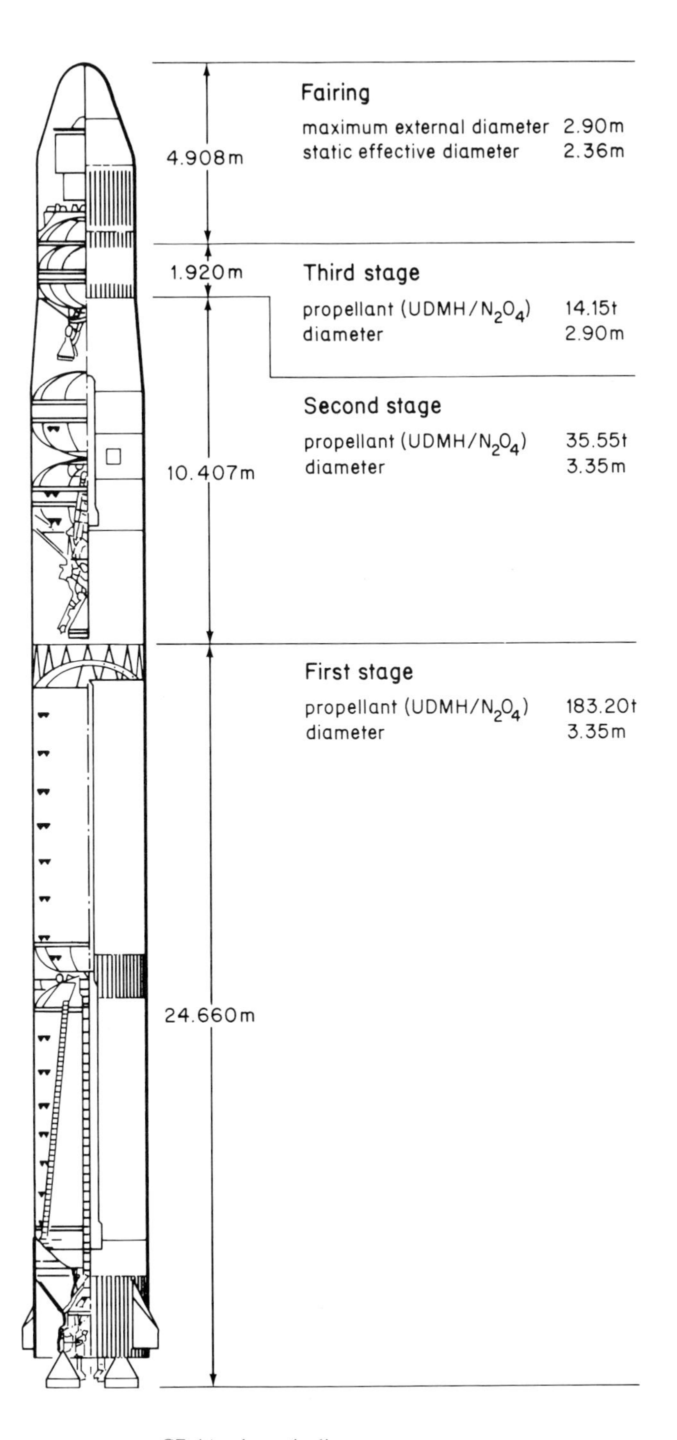

CZ-4A schematic diagram.

Long March 4 (CZ-4)

The Long March 4 was developed in the 1980s in order to fly meteorological satellites (the Feng Yun 1 series) into polar orbit from the new launch site of Taiyuan. Technically it is called the Long March 4A, since a new version, the Long March 4B, is in development. In effect, the Long March 4 is a stretched version of the Long March 2, but with a totally new third stage, with storable propellants and new engine (two YF-40s). For the CZ-4, Chinese rocket designers stretched the CZ-2C first stage by 4 m and the second stage by 3 m. The main difference in Long March 4B is that it has a more powerful, restartable third stage which is three times the length of the present third stage. Its thrust level is 3 per cent greater, and it can burn for a third as much longer. It would be able to lift 4.2 tonnes to low Earth orbit or 2.8 tonnes to polar orbit.

Table 6.12. Long March 4 launcher (CZ-4)

Dimensions	41.9 m long, 3.35 m diameter
Weight at lift-off	249.65 tonnes
Capability	1,500 kg to Sun-synchronous, polar orbit at 900 km; or 3,800 kg to 70° inclination, 400 km, low Earth orbit.

Flight record

6 Sep 1988	Feng Yun 1-1	757 kg	Meteorology.
3 Sep 1990	Feng Yun 1-2	881 kg	Meteorology.
	Qi Qui Weixing 1	4 kg	Small balloon to test atmospheric density.
	Qi Qui Weixing 2	4 kg	Small balloon to test atmospheric density.

	1st stage	2nd stage	3rd stage
Dimensions	24.66 m × 3.35 m	10.407 m × 3.35 m	1.92 m long × 2.9 m
Weight	192.2 tonnes	40.05 tonnes	15.15 tonnes
Engines	Four YF-20B	YF-22B	YF-40
Fuels	UDMH, nitrogen tetroxide	UDMH, nitrogen tetroxide	UDMH, nitrogen tetroxide
Thrust	302.8 tonnes	73.6 tonnes	10.2 tonnes
Burn time	156 s	127 s	5 min 21 s

Feng Bao-1 (1974–81)

Feng Bao was conceived around the same time as the Long March 1 and, like the CZ-1, benefited from the experience of the DF-5 missile. The rocket was developed in record time. The first attempt to launch the FB-1, on 14 July 1974, failed, but the rocket was successfully used to launch the Ji Shu Shiyan Weixing series of three technology payloads

(1975–6) and, following a further failure in July 1979, the three-in-one launch of Shi Jian 2, 2A and 2B in September 1981.

Because of its erratic launch record, coupled with the fact that the rocket had to rely increasingly on equipment taken from the Long March 1 and the availability of the Long March 2, it was felt that its continued existence was superfluous. The association of Shanghai with Mao Zedong's then-discredited political fortunes probably did not help. In September 1981, after the three-in-one launch, the staff in Shanghai were transferred to the Long March 3, whose development was then reaching a critical stage.

Table 6.13. Feng Bao launcher (FB1)

Dimensions	32.77 m long, 3.35 m diameter
Weight at lift-off	192 tonnes
Capability	1.9 tonnes to 200 km

Flight record

Date	Payload	Mass	Remarks
14 Jul 1974	(Ji Shu Shiyan Weixing)		(Second stage failed).
26 Jul 1975	Ji Shu Shiyan Weixing 1	1,100 kg	Technology satellite.
16 Dec 1975	Ji Shu Shiyan Weixing 2	1,100 kg	Technology satellite.
30 Aug 1975	Ji Shu Shiyan Weixing 3	1,100 kg	Technology satellite.
28 Jul 1979	(Shi Jian)	770 kg	Second stage failed.
19 Sep 1981	Shi Jian 2, 2A, 2B	257, 283, 28 kg	Three-in-one science mission.

	1st stage	2nd stage
Dimensions	17.835m by 2.25m	12m long
Engines	Four YF-20	Five YF-20
Fuels	UDMH, nitrogen tetroxide	UDMH, nitric acid
Thrust	280 tonnes	
Burn time	120 s	

SOUNDING ROCKETS

Our knowledge of the sounding rocket programme in China is incomplete: there is no full list of all the tests carried out since the programme began in the late 1950s. The role of the T-7 in preparing the way for the first Earth satellite has already been described, but it appears that other sounding rockets were fired during this period. Histories have some tantalising references: for example, small sounding rockets were tested from flat ships on inland lakes in 1964[36].

He Ping 2

Following the T-7 in the 1960s, two meteorological rockets were developed and used: He Ping ('peace') 2 and He Ping 6. The He Ping 2 two-stage solid fuel meteorological rocket was developed for the armed forces. Designer Song Zhongbao began work on the project in 1965 under the auspices of the China Academy of Sciences Applied Geophysics Research Institute. He Ping 2 was 6.645 m tall, weighed 331 kg and was able to reach 72 km. Its first flight took place in 1967, and serial production began the following year. He Ping used a new launch site – an undisclosed location in Heilongjiang in the north-west of the country. A total of 49 He Ping 2s were launched there from January 1970 to February 1973.

He Ping 6

He Ping 2's successor, He Ping 6, was much smaller, weighed only 60 kg, and was capable of firing a 2.8-kg package of instruments to 80 km. Considerable efficiencies were achieved by miniaturisation and by the development of a high-thrust engine. Made by the Space Physics Research Institute, He Ping 6 deployed a 1.6-m reflecting ball to act as a tracking target during its descent. He Ping 6 used a mobile launcher with guide rails, and on its first flight, from Jiuquan at the end of 1971, it reached 75 km. Four batches of tests were carried out between 1972 and 1975, but the results were disappointing due to poor quality equipment resulting from the Cultural Revolution. The final set of tests, made in Yunan in December 1979, were completely successful, sending nine rockets to an altitude of between 68 km and 90 km. A new sounding rocket, code-named project 761, was ordered in 1977, but its outcome is unclear.

Zhinui – Weaver Girl

Sounding rocket tests resumed with four Zhinui launches in 1988. The first Zhinui (Weaver Girl) launch took place in the morning of 19 December 1988 in the newly commissioned launching pad near Haikou, capital of China's southernmost island province of Hainan. At 2 min 10 s into the mission the first stage dropped away and later fell into Beibuwan Bay. The rocket reached such an altitude that the instruments took 2.5 hours to return to Earth, landing some 64 km distant from the launch site. A further three Zhinui 1 tests were made over the following six days to conclude the series. The main purpose of the flights appears to have been to collect information on the structure of the atmosphere[37].

On 22 January 1991, China began a new series of sounding rocket tests, sending the Zhinui 3 sounding rocket to an altitude of 120 km. The Zhinui 3 programme was begun by the Academy of Sciences in 1988, its purpose being to measure the elements and density of the atmosphere to an altitude of 120 km. The Zhinui rocket itself was 4.87 m tall, weighed 285 kg and had a payload of 45 kg. It took 3 min 24 s to reach the highest point of its ballistic arc.

ROCKET ENGINES

Sergei Korolev, Russia's great chief designer, once remarked that at the heart of a successful space programme lay a sound rocket engine. It may be no coincidence that rocket engine design was a high priority in the Russian programme, as the country has continuously been the leader in this field from the earliest days to the present time.

In Russia, rocket engines were designated RD- (raketa dgvatel), or rocket engine, from RD-1 onwards. China has followed a similar system, using the designator YF-, or yeti fadong (liquid type engine). Data on Chinese rocket engines are much less satisfactory than those on the Russian engines. In some cases, only one poor-quality photograph is available. The Chinese have not marketed their rocket engines in the West, but the Russians have – and with considerable success.

China has developed only a small number of rocket engine types, though they have built a number of variants. In essence, there are four: the YF-1 to YF-3 series, the YF-20 to YF-22 series, the YF-40 series, and the YF-73 and YF-75 series. These rocket engines have been adapted and modified to serve the entire range of the Long March and Feng Bao families. In addition, China has developed a small number of solid-rocket motors and minor engines.

YF-1 and YF-3 series

China's first liquid fuel engine was the YF-1 – the motor used for the Dong Feng 3 and Dong Feng 4 ballistic missile from the 1960s. Using UDMH as fuel, the YF-1 has a thrust of 28 tonnes. For the Long March 1, four YF-1s were clustered together to provide a thrust of over 100 tonnes (a configuration called the YF-1A).

The other early rocket motor was the YF-3, which was the engine for the second stage of the Dong Feng 4. It also uses UDMH. Design started in 1965, the first engine being completed 14 months later and the first test run being carried out four months after that. The YF-3 was designed to operate from an altitude of 60 km. For the Long March 1, a solid-fuel upper stage, the GF 02, was added.

Designers of the YF-1 and YF-3 engine

Ren Xinmin
Mo Tso-hsin
Zhang Guitian

YF-20 series

The YF-20 engine and its variants have been used for the Long March 2, 3 and 4 rockets, being introduced on the Long March 2 in 1975. It has a thrust of 70 tonnes and uses UDMH. Chief designer was Li Boyong. Design began in 1965, exhaustive tests were run over 1966–8, and by the following year the Chinese had clustered four YF-20s to provide a lift-off thrust of 280 tonnes (this configuration is called the YF-21). An improved version, the YF-20B, with 7 per cent more thrust, was developed for the Long March 2E, 3A, 3B and 4 (when clustered, these are called the YF-21B). The YF-20B was introduced on the Long March 2D in 1992. A single YF-20B engine is used on each strap-on booster for the Long March 2E and 3B.

The YF-22 engine, designed to ignite at altitude, is a modification of the YF-20, and is used for second-stage rockets. It was introduced on the Long March 2C second stage in 1975, and an improved version, the YF-22B, with extra thrust, was introduced on the Long March 2E second stage in 1990. The fuel used on both is UDMH.

YF-40 series

The YF-40 is the third-stage engine used on the Long March 4 rocket introduced in 1988. Third stage engines are relatively small in size and thrust compared with the first and second stage, but they have longer burn times, in the order of 320 s.

Liquid hydrogen third stage engines: YF-73 and YF-75

When Long March 3 flew in 1984, China became the third country in the world to master liquid hydrogen-fuelled upper stages. It was preceded by the United States (Centaur) and Europe (Ariane), and followed by the Soviet Union (Energiya). A powerful upper stage is highly desirable in order to put large comsats into geostationary orbit, but a complication is that the third stage must be restartable, firing once to enter Earth orbit, and a second time about 50 minutes later for the transfer to geostationary orbit.

Design of a third-stage restartable hydrogen-fuelled engine, the YF-73, began in 1965, the first initial hardware tests being carried out in 1971. Testing of the restarting capability began in 1979, and by the time it was ready for flight over 100 tests had been carried out. Despite this, however, development of the upper stage continued to prove problematical, and it failed to restart on its first flight in January 1984. This problem must have been promptly identified and remedied, as the next mission, four months later, went perfectly. The thrust of the liquid hydrogen third stage is 4.5 tonnes, with a burn time of 13.3 minutes. An improved version, the YF-75, with a thrust of 8 tonnes, was introduced with the Long March 3A in 1994. Restarting problems have, disappointingly, recurred.

Minor engines

The second stage of Long March 2 carries vernier engines which provide additional thrust and more stability at the bottom of the rocket. The engine motors concerned are the YF-23 (Long March 2C from 1975) and the YF-23B (Long March 2E from 1990). These low-thrust engines are used not just for steering but to put the second stage into orbit.

Solid rocket motor engines: GF-02 and EPKM.

China has developed two solid-rocket motor engines: the GF-02 and the PKM. (The acronym PKM denotes, somewhat crudely, perigee kick motor.) The two-stage Long March 1 rocket lacked sufficient thrust to reach orbit, and accordingly, a small upper stage was required. In developing a solid-rocket upper stage, the Chinese relied on solid rocket motors used on sounding rockets. The GF-02 was designed in 1967 at the Solid Fuel Engine Research Academy, the chief designer being Yang Nansheng, and was first tested the following year. The GF-02 exploded on its first test on 26 January 1968, but was subsequently mastered, 19 firings (including five altitude tests) having been carried out before its historic first assignment in April 1970. For the Long March 1D, the new light launcher, a new third solid-rocket motor engine, the F-G3b, has been developed.

The PKM was developed as a small kick motor to complete the transfer of comsats to geostationary orbit. There are two versions – the basic (PKM) and the EPKM (used on the Long March 2E). Built by the Haxi Chemical and Machinery Company, it is 1.7 m in diameter, 2.5 m long, and weighs 5,978 kg, of which 5,444 kg is propellant. On two Long March 2E missions, the Americans insisted on using their own solid-rocket fuel kick motor, the Star 63F. Both these missions were lost a minute after liftoff.

RELIABILITY

At the 1996 International Astronautical Congress in Beijing, the President of the Chinese Academy of Launcher Technology (CALT) declared that the primary problem for the Chinese space programme was trying to ensure the reliability of its launchers. At that time, Chinese launchers had become uninsurable. The main commercial launching base, Xi Chang, which had known a flurry of activity only a few years earlier, was idle, and no rockets were to be seen.

The development of the space shuttle by the United States in the 1980s began to create the public notion that somehow spaceflight could be made as reliable and safe as flying in an aeroplane. The loss of *Challenger* was a horrific reminder that this was not and could not be the case, and it is highly unlikely that rockets will ever achieve aeroplane-like reliability. From time to time, the major rocket programmes of all the major space powers suffer periodic reminders as to just how complex and dangerous modern rocketry remains. Even America's most reliable rocket, the Delta, was vulnerable, exploding soon after lift off in January 1997.

Despite this, the problems which the Chinese experienced with their Long March rockets in the 1990s went far beyond the normal limits of random failures. Statistically, their problem appeared to be worsening, rather than improving. For the period 1970–96, the Chinese overall reliability rate was 78 per cent. This fell to 50 per cent for 1995 and a dismal 33 per cent for 1996[38].

So, just how reliable are Chinese rockets? Table 6.14 lists the total number of known launches and failures (end of 1997). Many of the 1990s failures were highly publicised disasters flying high-profile missions: for example, Intelsat 708 in the 'Saint Valentine's Day Massacre'. However, it is neither fair nor accurate to state that as a general rule, Chinese rockets are unreliable. Some are extremely reliable. The Long March 1, 2C, 2D, 3A and 4 have 100% reliability rates.

A broader context is important. Three rockets were lost on their maiden flights: the Long March 2, Feng Bao and Long March 3B. Maiden flight losses are generally a third of all first-time launchings, so this problem is well within international norms. Furthermore, in all space programmes, there will be occasional random failures. Two missions were lost when the American Star 63F upper stage was used. It seems that there may be problems in fitting an American upper stage on a Chinese Long March launcher. Integration is the responsibility of both the Chinese and the Americans, and the blame should not be laid exclusively with the Chinese.

If the Chinese space programme is examined overall, in a 27-year context the reliability record (83.6 per cent) is within international norms for a developing space programme. When comparing Chinese *commercial* launchers – where the highest standards must be expected – the picture is mixed, both extremes being evident (100 per cent record for the Long March, 2C and 3A, and 66 per cent for the Long March 3B). The overall reliability record of China's *commercial* rocket fleet is 86.3 per cent. However, there still remains a

Table 6.14. Chinese launch attempts and failures, 1970–97

Launcher	Design bureau	First launch	No. of launches	No. of failures	Reliab-ility (%)
Long March 1	Chinese Academy of Launcher Technology	24 Apr 1970	2	0	100
Long March 2	Chinese Academy of Launcher Technology	5 Nov 1974	1	1	0
Long March 2C	Chinese Academy of Launcher Technology	26 Nov 1975	16	0	100
Long March 3	Both (CALT and SAST)	29 Jan 1984	12	3	75
Long March 4	Shanghai Academy of Space Technology	7 Sep 1988	2	0	100
Long March 2E	Chinese Academy of Launcher Technology	16 Jul 1990	7	2	71
Long March 2D	Shanghai Academy of Space Technology	9 Aug 1992	3	0	100
Long March 3A	Chinese Academy of Launcher Technology	8 Feb 1994	3	0	100
Long March 3B	Chinese Academy of Launcher Technology	14 Feb 1996	3	1	66
Feng Bao 1	Shanghai Academy of Space Technology	18 Sep 1973	6	2	33
			55	9	83.6%

string of recent upper stage failures associated with Long March 3 and 2E in which several satellites were stranded in incorrect orbits. Although failures to reach geostationary orbit can afflict the space programmes of other nations from time to time, the number of such failures has been higher than the international norm – and more than can be explained away by random failure. What has been most frustrating for the Chinese is that the causes appear to have been different and unrelated on each occasion.

Reliability figures are always problematical, for there are many ways of constructing them. Table 6.15 is an attempt to show Chinese reliability rates in a broad international context, and it shows how some Chinese rockets have much better reliability rates than Western ones. Overall, though, the Chinese record stands significantly lower than the best Western standards. It should be remembered, however, that the high-performing Western rockets have a much longer development history and have flown many more missions, giving their respective countries a much longer time to 'iron out the bugs' in their systems. Delta, for example, has been flying since 1960, Ariane has flown almost 100 times, and Proton several hundred times. Most failures occur early in a rocket's development programme, after which failures can become quite rare. China's commercial rockets are still in an early stage of development, and are comparatively little flown compared with some western launchers.

Table 6.15. World commercial launcher reliability rates (1997)

Launcher	Reliability rate (%)
Long March 2C, 2D	100
Long March 3A	100
Delta (United States)	95
Ariane (Europe)	92
Proton (Russia)	88
Atlas (United States)	86
Long March 3	75
Long March 2E	71
Long March 3B	66
All Chinese commercial rockets	86.3
All Chinese commercial rockets for geostationary orbit	76

Doubly frustrating is that these problems have emerged in a programme which has always had a strong commitment to quality control and testing. The Chinese do not claim to be world leaders in rocket and spaceflight technology; neither do they claim that their equipment is the most advanced in the world, for they know it is not. But they will not accept that it is inherently below professional standards. Because Chinese space budgets are restricted, the programme can afford exploding rockets and satellite breakdown much less than others. Each launch costs in the order of 10 million yuan. Accordingly, there is a strong emphasis on quality control and rigorous ground testing, considerable resources being so invested, as chapter 5's description of the extensive space infrastructure has indicated. The Chinese have introduced a 'testing pyramid' of checking individual components, combined parts and each system as a whole.

ASSESSMENT AND CONCLUSIONS

The relatively low annual rate of launchings in China is a function of economic factors rather than of technological shortcomings. China has an impressive family of launchers and variants, and is able to launch a range of payloads into several types of orbit, from small to large payloads into low Earth orbit, to specialised payloads into polar and Sun-synchronous orbits, and large communications satellites into geostationary orbit. Such versatility is the result of the imaginative adaptation of Long March to fulfil a variety of possible tasks. In a mirror image of the way in which Long March has been adapted, Chinese rocket engine technology shows the ability to adapt a limited range of models to suit a variety of launchers. Chinese rocket engines have demonstrated a high rate of reliability, and the failures which have arisen in the space programme have generally affected other parts of the rocket and its structure.

Whether or not China carries out this full range of missions in the rapidly expanding global space industry depends on a restoration of confidence in Chinese launchers. Detailed examination of the statistics of Chinese launchers shows that they are not inherently unreliable, as some ill-informed Western commentators have suggested. They do, however, highlight the much tougher and more costly learning curve for new space nations, and the fact that some individual launchers have particular problems which must be addressed. The Chinese realised from the start that they, of all countries, cannot afford failures, and from the beginning invested huge efforts in quality control and testing. Western observers of the Chinese space programme have commented that while the technological capacity of the programme may lack depth, standards of manufacture and quality control have been very high[39]. This experience is certain to hold them in good stead as they rebuild their troublesome commercial launchers.

China maintains three main cosmodromes. Whilst small in scale compared with Cape Canaveral, Vandenberg, Baikonour or Plesetsk, the three launch sites are able to carry out a flexible range of missions. Whilst some of the launch sites may have been basic, they have been modernised in recent years. Xi Chang is able to handle demanding payloads to the highest of standards. If, as we expect, Jiuquan is modernised to take the Long March 2E rocket in connection with the manned space flight programme, it may become one of the most modern in the world and be much better known than it is now. It is to this topic – the manned space flight programme – that we now turn.

7

China's future in space

By the late 1980s the Chinese space programme had succeeded in developing scientific, recoverable, applications and communications satellites and launching them using an indigenous fleet of booster rockets. They had achieved all this independently, with little or no assistance from or cooperation with other nations.

Perhaps the moment which marked China's coming of age in the international space community was the 47th International Astronautical Federation Congress, held in Beijing in October 1996. Attended by 2,000 domestic and more than 1,000 foreign delegates, the IAF congress was opened by Chinese president Jian Zemin and hosted by prime minister Li Peng[40]. The event might have taken place in Beijing sooner had the events of Tiananmen Square not taken place, and sceptics pointed to the exercise as an attempt by the Chinese to mend their political fences with the West. The Chinese presented their space industry on show, took westerners around Chinese space facilities, unveiled plans for their own space future and appealed for greater international cooperation between China and its international partners.

This chapter reviews the new projects planned for the last years of the twentieth century and the early years of the next century, and describes China's progress towards a manned space programme.

MORE AMBITIOUS PLANS: THE MANNED SPACE PROGRAMME

Although the Chinese have had the launching capacity to organise a much wider range of military and science applications and other programmes, insufficient resources have prevented the Chinese from doing so. They have concentrated on quite a narrow range of mission types – recoverable cabins, weather satellites, comsats and a small number of scientific missions. China has launched no missions to the Moon or planets, even though countries which began to launch satellites at a similar time have now done so (Japan sent its first spacecraft into lunar orbit in 1990).

The Chinese, like everyone, were greatly impressed with Yuri Gagarin's historic flight into space on 12 April 1961, and the Academy of Sciences was spurred into holding a series of symposia starting that summer. Tsien Hsue-shen was one of the organisers. By 1964, twelve meetings had been held, their purpose being to keep in touch with developments abroad and discuss how a manned and deep space exploration programme could best be organised in the distant future.

China's first work in space medicine dates to the mid-1960s when the first space-related medical experiments were carried out by the Academy of Military Medical Sciences and the Chinese Academy of Medical Sciences. Such tests were approved in the guidelines for the exploration of space approved by the Central Committee in August 1965, when it approved the construction of the Earth satellite.

FIRST RUMOURS OF MANNED SPACEFLIGHT

Historically, the greatest mystery within the Chinese space programme has been the manned programme. The first rumours of a manned Chinese space flight programme can be traced to the start of the FSW missions in 1975[41]. The FSW cabins, being comparable in size with the American Mercury missions of 1961–63, were (only just) large enough to contain a man. As if to give credence to speculation about the purpose of the FSW missions, the Chinese then published photographs of what appeared to be cosmonauts in spacesuits undergoing training in simulators, 25-m drop chairs, centrifuges, altitude chambers, vibration systems, spinning gondolas and aircraft (to simulate zero gravity). Some were even settling down to a hearty meal of space food[42]. The candidate cosmonauts trained in a mockup spacecraft which could simulate high altitudes and there were even star views outside the windows to test the ability of the trainees to recognise the constellations. The men concerned were identified as young pilots from the Chinese air force. (The Russians had published similar photos in the late 1950s, provoking a similar flurry of rumours.) But despite the speculation,the moment passed.

In reality, the idea of a manned space flight had always been in Chinese thinking from the very start. Tsien's book *An introduction to interplanetary flight* – the basis for instruction of all engineers in the space programme – included a chapter on manned space flight. On 1 April 1968 he set up the Research Centre into Physiological Reactions in Space. Its existence was not revealed for 20 years, by which time it had carried out 455 successful experiments to predict how humans would react to space travel. The centre was equipped with acceleration chairs, pressure chambers, centrifuges and revolving chairs. It even investigated whether traditional Chinese breathing exercises – qi gong – could be of help to cosmonauts in space[43]. It is probable that these tests were strictly medical experiments, but it is easy to interpret them as more active advance preparations for a real space flight.

How serious were these preparations? It is difficult to judge. In January 1981 the general secretary of the Chinese Space Research Society, Wang Zhuanshan, was quoted as saying that China had postponed a manned spaceflight for at least ten years due both to economic considerations and a reappraisal of Chinese space aims and objectives. But his remarks of themselves suggest that a manned flight had at one stage been considered more actively and much sooner[44].

In 1978 the first non-Russians and non-Americans began to fly in space. The first, a Czech, was flown by the Russians up to the Salyut 6 manned orbital station. Other Soviet-block countries followed. With the introduction of the shuttle in 1981, a range of other nationalities began to fly in orbit. President Reagan, in the course of a visit to China in spring 1984, offered the Chinese a seat on the American space shuttle, an offer which had to be withdrawn after the *Challenger* disaster in January 1986. Mikhail Gorbachev also offered the Chinese a seat into space – to the Mir space station – but this also did not progress far. It seemed that China preferred to settle for the longer, tougher and more demanding task of becoming the third country to put up its own cosmonauts through its own efforts.

NEW ROUND OF RUMOURS

In the mid- to late 1980s, the rumour mill swung into action again, and further photographs of fresh training candidate cosmonauts appeared. These were reinforced by the fact that the Long March 2E was in development: a booster capable of putting 9 tonnes into low Earth orbit, more than enough for a manned spacecraft, and heavier than the Russian three-man Soyuz. *China Daily*, carrying a story from the *People's Daily*, reported in September 1986 that China had made a simulated space cabin, that selection of an astronaut group was under way and that a life support system had been fully tested[45]. In summer 1992 a Hong Kong paper reported that a manned spaceflight centre was being built near Jiuquan to support a manned space programme, and in August reports were carried that China would launch cosmonauts by the year 2000.

At the International Astronautical Federation Congress in 1992, Ziandong Bao presented a paper entitled *A modular space transportation system*, which outlined a new launcher system able to loft 11 tonnes into low Earth orbit – more than adequate for a small manned spacecraft. Two years later the China Aerospace Corporation (CASC) issued a brochure of a modular Chinese space station, showing four Soyuz-class modules with solar arrays clustered around a central hub. The picture was reminiscent of early 1970s Russian designs.

RENEWAL OF CONTACT WITH MOSCOW

The next move took the Chinese back to their long-estranged partners in Moscow. Links were reopened in early 1993 when the chief of staff of the People's Liberation Army, Chi Haotian, visited Star Town, the cosmonaut training centre in Moscow. Formal agreement on cooperation between the Russian Space Agency (RKA) and the Chinese National Space Administration was signed in Moscow on 25 March 1994.

In March 1995, a group of Chinese space experts visited Moscow for the first time since the great split in 1960. They expressed an interest in buying equipment which could be used for manned space flight, such as environmental control systems, docking and emergency systems, and bought some RD-120 rocket engines, the advanced engines which the Russians use on the second stage of their Zenith rockets. They tried to buy the world's most advanced first stage engine, the RD-170, but the Russians declined to negotiate. Russian experts made a return visit to China later that year, after which the Chinese bought an entire spacecraft life support system. The Chinese also bought an Energiya docking module and the Kurs rendezvous system, used to dock supply craft with the Mir space station.

On 20 August 1996 a number of Chinese (estimates of numbers ranged from 20 to 50), led by Shen Jungjun, arrived in Star Town. Although Star Town had traditionally been

Li Tsinlung. Photo courtesy Neil Da Costa.

very much off limits to Westerners, this had ceased to be the case when the collaborative missions with the Americans began in early 1995; indeed, the centre was crawling with American specialists and astronauts. Although the Russians were unusually coy about the presence of the Chinese in Star Town, they could not deny that they were there, for they could not exactly by hidden.

So what were the Chinese doing in Star Town? There were many conflicting interpretations of their presence. Two of them – Wu Tse and Li Tsinlung – were named in December 1996, and were variously described as cosmonauts, cosmonaut trainees and instructors. The most reliable interpretation of their role is that they were cosmonaut instructors who would not be flying into orbit themselves. After a year, they would return to China and train in the first cadre of Chinese cosmonauts. Apparently, the Chinese paid $1 million for the training and medical advice and supervision of space training.

There are several reports that the Chinese already have their own manned space training facility. In 1988, unconfirmed reports were published of a cosmonaut training centre located in a walled village in the western suburbs of Beijing and described as well up to the standards of facilities available in Moscow and Houston. The centrifuge was the biggest in the world, with a rotary arm of 30 m[46].

PROJECT 921

The nature of Chinese intentions was publicly (though incompletely) revealed at the 1996 Beijing conference[47]. The vice-administrator of the China National Space Administration, Wang Liheng, announced that China intended to make a breakthrough to manned space flight before the end of the century. It was revealed that a manned space programme had been decided in 1992, and was code-named project 921. This, indeed, would fit the known Chinese practice of assigning a three-digit code to approved projects. An unmanned prototype would fly in 1998, with a manned flight by two cosmonauts in a Gemini-class cabin before October 1999, just in time to mark the 50th anniversary of the Cultural Revolution. China's ultimate intention was to have its own manned space station by 2020.

At the Paris Air Show in 1997, the Chinese delegation confirmed that work was proceeding on the Chinese manned space flight, and that work had already started on the new Long March pad at Jiuquan. The main effort at present was being concentrated on the adaptation that would be necessary for the likely launcher, the Long March 2E. Considerable efforts would be made to improve its electronics, guidance and quality control.

Wu Tse. Photo courtesy Neil Da Costa.

CHINESE SPACE SHUTTLE

To confuse the picture, however, the Chinese also spoke in terms of a 5-man spaceplane – suggesting that they would skip the intermediate stage of a Gemini-class spacecraft – and gave a launch date of 2000 for two demonstrator flights, with the first manned flight, by two cosmonauts, in 2001 or 2002. The Chinese have never made any secret of their interest in shuttle designs.

The first Chinese studies of a space shuttle had been presented much earlier, at the 1983 International Astronautical Federation congress in Budapest, Hungary, when Zu Huang of

Li Tsinlung and Wu Tse in training in Star Town, Moscow. Photo courtesy of Neil Da Costa.

the Chinese Academy of Sciences presented *Fully reusable launch vehicle with airbreathing booster*, the design of a two-stage ramjet-powered shuttle along the lines first proposed by the Austrian scientists Eugen Sänger and Irene Bredt in the early 1940s. At the 1990 conference of the International Astronautical Federation, Shusheng Wang and Kexun Zhang presented another feasibility study of a Chinese space shuttle on behalf of the Third Institute of Research in Beijing. The bottom stage is 45 m long, with a 198-tonne reusable carrier vehicle using tri-propellants (liquid oxygen, hydrogen and methane) and ramjets. At Mach 6.5, the 132 tonne, 35 m long, 15 m wingspan orbiter is released to continue its path toward orbit while the mother craft returns to Earth like a jet plane. The shuttle design is much smaller than the American orbiter or the Soviet Buran, and is closer to two European concepts – Germany's Sänger and the French Hermes. It could fly three cosmonauts or bring 6 tonnes of cargo to an altitude of 500 km.

This, therefore, represents what is known of the Chinese manned space programme. China has tested many of the key technologies necessary for manned space flight, such as the recovery of space craft from orbit (indeed, that is one of the most successful parts of their programme), and they have several powerful launchers able to launch manned spacecraft. Key parts of the infrastructure – a school of space medicine, Jiuquan launch site, cosmonaut training facilities – are already in place.

The Chinese, however, have been much less forthcoming about the spacecraft design they intend to follow, and it is unclear if it will be comparable with the American Gemini of the 1960s or the venerable Soviet Soyuz, or will even be a small spaceplane. Neither have they stated whether manned flights are intended to support a manned space station programme either immediately or in the long-term. They have set about the project in their typical manner: thorough long-term background preparation, the acquisition of experience from other countries, the training of trainers, and the eventual designation of a project code. When they are ready, they will fly.

NEW LAUNCHERS

What are China's other space plans? China intends to develop, probably early in the new century, a new class of launch vehicle, much more powerful than any of the existing Long March system, which may well be called the Long March 5. It will use liquid hydrogen for its main stage, much like the Americans mastered on the Saturn V, the Russians on Energiya and which the Europeans use on Ariane 5. The intention is to build a rocket that can put a 20-tonne spacecraft into low Earth orbit, or about 4 tonnes to geostationary orbit. Such a rocket would be well able to launch space station modules. This would require a large, powerful engine, the current standard being the Russian RD-170 used on the Energiya rocket. The Chinese are already working on a much more powerful rocket engine with conventional fuels, possibly developed from the RD-120 of the Russian Zenith.

Xiandong Bao's 1992 paper, *A modular space transportation system*, outlined a new family of launch vehicles. These were probably paper studies only, and indicative of Chinese thinking rather than definite intentions. The baseline vehicle was a large rocket of 377 tonnes with a payload capacity of 11 tonnes to low Earth orbit from Jiuquan; or, with a third stage, 6 tonnes to geostationary transfer orbit. Both the second and third stages would use liquid hydrogen as a fuel. Perhaps the most interesting feature of the proposal was that the first stage would use non-storable propellant (probably kerosene), the first time since the 1960s that the Chinese have used non-storable fuels.

Already available is Long March 1D, a new light booster able to put 150–1,000 kg into orbit. Despite the need for a substantial number of rockets able to put large payloads into orbit in the 1990s, there has been a huge growth in the requirement for small payloads in low Earth orbit. An American company, Orbital Sciences Corporation, responded by developing the Pegasus, a small rocket lifted to altitude by a Lockheed Tristar, whence it was dropped, ignited and flew on to orbit. The Russians developed a range of small rocket launchers (Rokot, Start), and the Americans the Athena. The Chinese, for their part, have turned back to Long March 1, which has not flown since 1971.

The Long March 1D is smaller and lighter, but more powerful, than the Long March 1. It has new engines (YF-2A, two YF-40 and FG-6 on the three stages respectively), is slightly lighter (81 tonnes) and has a much-improved performance – able to put up a 750 kg satellite. In addition, the CZ-1D offers improved electronics and guidance systems taken from the CZ-3. The promoters, China Great Wall Industry Corporation, offer both low Earth, Sun-synchronous or polar orbits. Although the rocket has been available since 1991, it has still not found any customers. The initial price on offer was £10 million ($16 million).

A new heavy rocket, Long March 3C, also awaits customers. This has two solid rocket booster strap-ons, rather than the four of the 3B, and is able to place 3.5 tonnes in geostationary orbit.

NEW SPACECRAFT PROGRAMMES

The current framework for the Chinese space programme may be found in the *National long and medium-term programme for science and technology development, 2000–2020*, This was prepared in the context of the ninth five-year plan agreed at the eighth National People's Congress, Fourth Plenary, of March 1996. The framework for the space programme emphasises the importance of applied satellites and commercial competition. The key elements of the 20-year plan are the development of comsats, metsats, satellites for remote sensing and other applications, providing international launcher services at competitive prices and a new launcher capable of putting 20 tonnes into orbit. China has a range of new spacecraft programmes at various stages of development.

An imaging radar satellite

Two imaging projects are under way. One is for a radar imaging satellite, perhaps modelled on the Russian Almaz or the more recent and more successful Canadian radarsat. China has plans to buy the imaging device used on Canada's radarsat and then adapt it to one of its own satellites. In September 1997, a receiving station was set up to pick up data from Canada's radarsat.

The other project is for a small Earth imaging satellite, sometimes called the Small Satellite Remote Sensing System. Developed by the Chinese Society of Astronautics and the Institute of Technology in Harbin, the proposal is for a 250 kg satellite with a telescope-based CCD able to scan images to a resolution of 1.5 m from 630 km. It is currently at project definition stage.

The start of planetary science

The Chinese would like to fly their first deep space mission early in the 21st century. So far, preliminary choices must be made about targets before spacecraft design can even begin. At the 1995 Chinese Academy of Sciences conference, the director of space research, Prof. Jiang Jingshan, said that a pre-study of a lunar satellite was under way. If approved, it could fly to the Moon in the year 2000. Paper studies have already been carried out by younger technicians with the Chinese Academy of Space Technology.

New small satellites

A significant trend in the world space industry in the 1990s was the move towards much smaller satellites, defined as payloads of less than 300 kg. A total of 45 such satellites were launched by other countries during the 1991–5 period alone. Partly, this was driven by the desire to reduce costs, but also, partly, made possible by great advances in computers, materials and miniaturisation. Shi Jian 4 has already provided the basis for China's development of small satellites.

China hosted the first Asia–Pacific workshop on multinational cooperation in space technology and applications in Beijing in 1993, and this afterwards became an annual event for the region in which the development of small satellites was the main focus. In 1996, Zhou Sumin, of the China State Seismological Bureau, announced that China was planning two 50-kg seismic relay satellites. Each satellite would operate in a 1,000-km orbit at 60°, picking up information on seismic activity and potential earthquakes as they passed over 800 instrument stations attached to 450 ground stations. They would store the information as they picked up the signals, and then quickly relay it to the ground as they passed over the national seismic monitoring centre in Beijing. This store-dump technique was developed by Russian spy satellites in the 1960s, and was used by China's own Shi Jian 2 in the 1980s.

A solar observatory

China is planning a 2-tonne solar observatory in 50/50 cooperation with Germany. According to Ai Guoxiang, academician of the Chinese Academy of Sciences, this is a $100 million project which involves the launching of a solar observatory in 2002. The telescope itself will weigh 1.2 tonnes and will have an angle of resolution similar to that of the American Hubble Space Telescope. Preliminary designs for the telescope will be tested on hot air balloons flying to 32 km in 1998.

Navigation satellites

In the late 1980s there were reports – which were probably untrue – that China had planned navigation satellites. Now, however, navigation satellites are under consideration. The proposal has sometimes been associated with the name TwinStar.

ASSESSMENT AND CONCLUSIONS

Manned space flight is the next big challenge for the Chinese space programme. It is clear that the Chinese have given consideration to manned flight since the 1960s and carried out a considerable amount of preparatory work, especially in the field of aviation medicine; however, there is insufficient evidence to believe that they ever came close to the design stage for a manned spacecraft.

From the early 1990s the perspective began to change, and the Chinese now have considerable expertise in launching sophisticated, powerful boosters, in running tracking systems, in operating launch sites and a broad range of space facilities, and in bringing back recoverable spacecraft. An entire generation of engineers has grown up in the space industry. The evidence of Chinese intent on manned spaceflight is much stronger now, as may be seen in the allocation of a project code (921), which is always an indication of seriousness in planning. Confirmation of China's sense of purpose may also be seen in the purchase of equipment from Russia, the setting up of new cosmonaut training facilities and the cooperation arrangements in Star Town in Moscow.

For the Chinese, manned space-flight would be a logical advancement for a country familiar with taking big steps: the Earth satellite in the 1960s, the recoverable cabin in the 1970s, the communications satellite in the 1980s – maybe manned flight before the end of the 1990s. One day, cosmonauts flying eastwards from Jiuquan will orbit the Earth and watch the yellow, orange and red glow of dawn march over the skies of China; and they will be able to see, with their own eyes, that the East is Red – *Dong Fang Hong*!

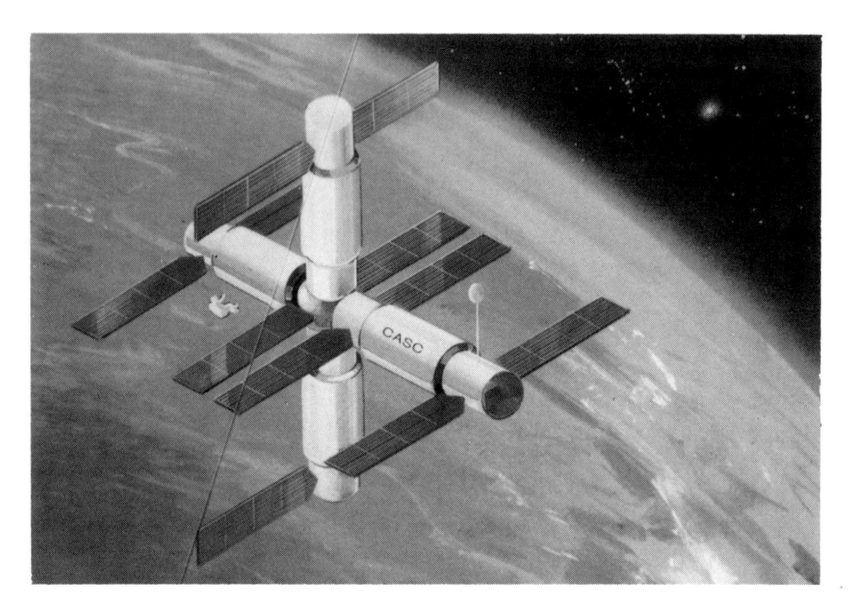

Chinese design for a manned space station, twenty-first century.

References

1 Basic Books, New York, 1995.
2 See this author. *The new Russian space programme - from competition to collaboration*. 2nd edition, Wiley-Praxis, 1996.
3 Zhu Yilin. 'Development of Chinese satellites under Prof. Tsien'. *Journal of the British Interplanetary Society*, **50**, 185–188, 1997.
4 Zhang Jun (*ed.*) *The Chinese space industry today*. China Social Sciences Publishing Co., Beijing, 1986.
5 Kenneth Gatland. Missiles and rockets. Blandford, London, 1975, pp. 216–221.
6 Joseph Anselmo. US eyes China missile threat. *Aviation Week & Space Technology*, 21 October 1996.
7 China said to be developing missiles that could hit Rockies. *International Herald Tribune*, 24 May 1997.
8 Joseph Anselmo. China's military seeks great leap forward. *Aviation Week & Space Technology*, 12 May 1997.
9 Zhang Jun. *The Chinese space industry today*. China Social Science Publishing Co., Beijing, 1986.
10 Zhang Jun. *The Chinese space industry today*. China Social Sciences Publications, Beijing, 1986.
11 Chen Heyi (*ed.*). *Into outer space*. China Pictorial Publishing Co., 1989, p. 156.
12 James Harford. *Korolev*. Wiley, New York, 1997.
13 Zhang Jun. *The Chinese space industry today*. China Social Science Publications, Beijing, 1986.
14 Zhang Jun. *The Chinese space industry today*. China Social Science Publications, Beijing, 1986.
15 Zhang Jun. *The Chinese space industry today*. China Social Science Publications, Beijing, 1986.
16. Zhang Jun. *The Chinese space industry today*. China Social Science Publications, Beijing, 1986.
17. Zhang Jun. *The Chinese space industry today*. China Social Science Publications, Beijing, 1986. Later the history refers to other satellite projects being abandoned in mid-course. However, some of them were not urgently needed, and it was said that the

preliminary work was a wasted investment. It would be interesting to know more of these projects.

18 The main source of information on China's recoverable satellite programme is Phillip S. Clark. *China's recoverable satellite programme.* Heston, Middlesex, Molniya Space Consultancy, 1994.

19 Chen Heyi (*ed.*). *Into outer space.* China Pictorial, Beijing, 1989, p. 43.

20 Zhang Jun. *The Chinese space industry today.* China Social Science Publications, Beijing, 1986.

21 For example, 'How creative can you be in marketing your products? Now, the sky's the limit!' *Aviation Week & Space Technology*, 15 March 1993.

22 'China faces commercial launch ban'. *Spaceflight*, **32**, May 1990.

23 James Asker. 'China, US renew Long March pact'. *Aviation Week & Space Technology*, 6 February 1995.

24 The best summary of these issues may be found in Lawrence H. Stern and Jack High. 'America takes a long march into space'. *Spaceflight*, **32**, April 1990.

25 Phillip S. Clark: *The Chinese space programme – an overview.* Whitton, Middlesex, Molniya Space Consultancy, 1996, pp. 124–127.

26 Craig Covault. 'Chinese Long March faces crucial return to flight'. *Aviation Week & Space Technology*, 24 March 1997.

27 Zhu Yilin. 'Space microgravity scientific experiments in China'. *Spaceflight*, **35**, October 1993.

28 Zhu Yilin. *Some results of Chinese space microgravity life science experiments.* Monograph.

29 Roisin Ingle. 'Satellite now set to land on Limerick'. *Sunday Tribune*, 11 February 1996. The claim that the heat shield was made of wood was made in 'In orbit', in *Spaceflight News* (October 1988), pp. 28–29 and in other reputable journals, but this does not appear to be borne out in the technical descriptions of the design of the spacecraft published by Clark, where the heat shield is described as made of molybdenum alloys and other composite materials (*op cit.*). None of the Chinese literature suggests the use of wood in the FSW series. The 'wooden heat shields' story may have its origin in a poor translation.

30 Zhang Jun. *The Chinese space industry today.* China Social Science Publications, Beijing, 1986.

31 The most recent account of the Chinese space organisational framework may be found in Christian Lardier. 'Les ambitions de l'industrie spatiale chinoise'. *Air & Cosmos*, 25 Octobre 1996.

32 The first detailed description of these launch sites was provided in Kenneth Gatland (*ed.*). *The Illustrated Encyclopaedia of Space Technology.* 2nd edition, Salamander, London & New York, 1989.

33 Christian Lardier: 'Première visit au champ de tir de Xi Chang'. *Air & Cosmos*, 25 Octobre 1996.

34 Phillip S. Clark. *Chinese launch vehicles.* Whitton, Middlesex, Molniya Space Consultancy, 1996.

35 In October 1960, a Soviet R-16 missile exploded at Baikonour cosmodrome. A total of 97 engineers, supervisors and rocket troops died in the ensuing fireball, but the

level of casualties was made much worse by the toxic nature of the exploding fuel. It remains the worst launch disaster in history.

36 Theo Pirard. 'Chinese secrets orbiting the Earth'. *Spaceflight*, **19**, (10), October 1977.

37 The most detailed accounts of these missions may be found in Phillip S. Clark. *Chinese space activity, 1987–8*. Astro Info Service Publications, West Midlands, 1989.

38 For more detailed research on reliability rates, see Pierre Lengereux and Christian Lardier: 'La Chine veut améliorer la fiabilité des ses fusées'. *Air & Cosmos*, 18 Octobre 1996.

39 For a detailed treatment of these issues, see the report by Harford and Pritchard on the American visit of Chinese space facilities in 1979.

40 Craig Covault. 'China seeks cooperation, airs new space strategy'. *Aviation Week & Space Technology*. 14 October 1996.

41 The most up-to-date review of the Chinese manned space effort may be found in Phillip S. Clark: 'Chinese designs on the race for space'. *Jane's Intelligence Review*, **9**, (4), April 1997.

42 Chinese astronauts train in simulators'. *Aviation Week & Space Technology*, 26 January 1981.

43 Yu Qingtian. 'Keeping fit in space'. *China pictorial*, November 1988.

44 For an exploration of China's space capabilities at this time, see Gerald Borrowman: 'China's long march to orbit'. *Spaceflight*, **25**, (5), May 1983.

45 This story is discussed in more detail in Phillip S. Clark: 'In business and advancing fast – Chinese space activity'. *Spaceflight*, **29**, February 1987.

46 G. Lynwood May: 'China advances in space'. *Spaceflight*, **30**, November 1988.

47 The principal Western report on this aspect of the conference was filed by Craig Covault: 'Chinese manned flight set for 1999 liftoff'. *Aviation Week & Space Technology*, 21 October 1996.

Principal milestones in the development of the Chinese space programme

790	Invention of the rocket in China by Feng Jishen.
1083	Rockets used by the Song dynasty to fight Xixia.
1275	Kublai Khan uses rockets to drive out the Japanese.
16th C	Wan Hu develops a two-stage fire rocket to ascend into heaven, assisted by kites. Naval rocket, the Fiery dragon.
1911	Birth of Tsien Hsue-shen in Hangzhou.
1926	Liquid fuel rocket fired in United States (Goddard).
1933	Liquid fuel rocket fired in Soviet Union (Korolev).
1935	Tsien Hsue-shen goes to study in the United States.
1937	Tsien Hsue-shen's first writings on modern rocketry.
1945	Tsien in Germany to survey German wartime achievements in rocketry. Arthur C. Clarke proposes the establishment of satellites in 24-hour orbit to provide global communications.
1947	Soviet Union fires German wartime rockets.
1949	Revolution in China.
1955	Tsien Hsue-shen repatriated to China.

1956	Jan	Establishment of Scientific Planning Commission in China.
		Adoption by China of *Long-range planning essentials for scientific and technological development, 1956–67*, committing China to the development of rocket and jet technologies.
	Apr	Central Committee invites Tsien Hsue-shen to outline the potential of guided missiles and rockets.
		Appointment of State Aeronautics Industry Commission
	Sep	Visit by Nie Rongzhen to Soviet Union.
	Oct 8	Central Committee of the Communist Party of China, presided by Mao Zedong, establishes the Fifth Academy to develop the space effort; the founding date of the Chinese space programme.
		Arrival of R-1 missiles from Soviet Union.
1957	Jul	Nie Rongzhen goes to Moscow a second time to ask for more advanced missiles.
	Aug 20	*New Defence Technical Accord 1957–87* signed with Soviet Union
	Oct 4	First satellite launched into Earth orbit by the Soviet Union (Sputnik 1). Chinese Academy of Sciences sets up seven observing stations.
1958	Jan	Fifth institute adopts the *Essentials of a ten-year plan for jet and rocket technology, 1958–67.* Adoption of project 1059 – to copy and fire a Russian R-2.
		Great leap forward.
		Russian R-2 arrives in China.
	Apr	20th corps leaves for Jiuquan to begin construction of China's first launch site.
	May 15	Soviet Union launches large scientific satellite (Sputnik 3).
	May 17	Mao Zedong declares: 'We too must launch artificial satellites'.
	Aug	State Scientific Planning Commission endorses proposal for China to launch an Earth satellite; proposal codenamed project 581.
	Nov	Task of construction of satellite given to Shanghai Institute of Machine and Electricity Design (SIMED).
1959	Jan	Plan to launch an Earth satellite shelved by Deng Xiaoping.
1960	Feb 19	Launch from Laogang of T-M prototype sounding rocket.
	Aug	The great split: Soviet scientists ordered home.
	Sep	Flight of T-7 sounding rocket.
	Sep	China fires Soviet-made R-2.
	Nov 5	China launches the first modern rocket manufactured in China, a copied R-2 (Dong Feng 1).
1961		Establishment of Shanghai Academy of Space Technology (SAST).
	Apr 12	First man in space, Yuri Gagarin (Soviet Union).

1962	Mar 21	Dong Feng 2 missile crashes at Jiuquan.
		SIMED team goes to Beijing to study in detail the idea of an Earth satellite.
1963	Jan	SIMED made part of the Fifth Academy.
	May	Start of interplanetary flight symposia.
	Dec	Flight of T-7A sounding rocket.
		Tsien Hsue-shen publishes *An introduction to interplanetary flight.*
1964	May	Academy of Sciences begins feasibility study into Earth satellite (concluded a year later).
	Jun 29	First successful flight of Dong Feng 2 missile.
	Jul 19	Flight of T-7A biological sounding rocket.
	Nov 23	Fifth Academy replaced by the Seventh Ministry.
1965	Jan	Tsien proposes an Earth satellite to the Central Committee.
	Mar	Preliminary work begins on liquid hydrogen powered rocket at Liquid Fuel Rocket Engine Research Institute; start of work on the YF-73 rocket engine.
	Apr	Tsien's proposal receives support from the party's Defence, Science and Technical Commission, recommending a satellite launch in the 1970–71 time period.
	Jul	Academy of Sciences proposes *A proposal on the plan and programme of development work of our artificial satellites.*
	Aug 10	Approval of the plan at a meeting chaired by Zhou Enlai.
	Aug	Allocation of project to institute 651 in Beijing.
	Aug	Decision to disperse production to third-line areas.
		Start of design of Dong Feng 4 and Long March 1.
	Oct 20	Start of 64-day conference on the design of the artificial Earth satellite, project 651.
		Start of construction of rocket engine testing site in Beijing.
1966	Mar	Cultural Revolution.
	May	Earth satellite named Dong Fang Hong; design revised.
		Introduction of the chief designer system.
	Jul 14	First Chinese space dog, Xiao Bao, flies into space on T-7A-S2 sounding rocket (second flight on 28 July, carrying Shan Shan).
	Oct 27	Dong Feng 2A used for live nuclear rocket test.
	Dec 26	First flight of the Dong Feng 3 missile.
		Start of conceptual studies for recoverable satellite programme.
		Start of construction of China Satellite Launch and Tracking General Control (CLTC).
1967		First flight of He Ping 2 sounding rocket from Heilongjiang.

	Mar 17	Zhou Enlai orders protection of key space workers from the Cultural Revolution, puts space programme under military authority.
	Jun 23	Setting up of mission control centre, Xian.
	Sep 11	Three-day conference on recoverable satellite programme (project 911).
	Nov	Establishment of the Chinese Academy of Launcher Technology (CALT), replacing the Beijing Wan Yuang Industrial Corporation.
1968	Feb 20	Establishment of Chinese Academy of Space Technology (CAST).
	Apr 1	Setting up of research centre into physiological reactions in space.
	Aug 8	Testing of solid rocket motor GF-01 on sounding rocket (again on 20 Aug).
1969	Jul 20	United States lands two men on the Moon.
1970	Jan 30	First flight of the Dong Feng 4 missile.
	Apr 24	China launches its first Earth satellite, Dong Fang Hong.
	May	Conference held to prepare the launch of China's first scientific satellite, Shi Jian.
	Aug	Military Commission, under Lin Biao, adopts five year plan for the launching of 14 satellites and eight new launchers over 1971–76.
		First project conference on a communications satellite.
1971		First flight of He Ping 6 sounding rocket from Jiuquan.
	Mar 3	China's first scientific satellite, Shi Jian 1, which operates for eight years.
	Sep	Lin Biao flees China for Soviet Union; plane shot down in flames.
	Sep 10	First flight of Dong Feng 5 intercontinental ballistic missile from Jiuquan.
1972	Aug 10	Sub-orbital test of the Feng Bao rocket.
		Visit of President Nixon to China.
1974	Nov 5	First attempt to launch Long March 2 (unsuccessful).
	Nov 25	Report on developing satellite communications in China is commissioned by the Central Committee.
1975	Feb 17	Report *Concerning the question of development this country's satellite communications* is presented.
	Apr	Mao Zedong gives go-ahead for communications satellite proposal (project 331).
	Jul 26	Launch of Ji Shu Shiyan Weixing 1, first of series of three satellites, probably military (project 701).

Year	Date	Event
	Nov 26	Launch of first recoverable satellite, the Fanhui Shi Weixing, recovered after three days. China is the third space power to recover satellites, after the United States and the Soviet Union.
1976	Apr	Death of Zhou Enlai.
	Aug	Decision to use liquid hydrogen for third stage of new rocket to launch communications satellites.
	Sep	Death of Mao Zedong.
	Oct	Military coup: crushing of the Gang of Four.
1977	Mar	Stations for communications satellites in geostationary orbit requested from International Telecommunications Union.
	Sep	Commissioning of comships *Yuan Wang 1* and *2*.
	Oct	Project conference on new communications satellite.
		Third plenum of 11th Central Committee adopts new plan for the exploration of space, involving Dong Feng 5, submarine-launched missile and communications satellites.
		Start of policy of cooperation with other countries.
1978	Aug	Deng Xiaoping says China will not take part in a space race but will devote its efforts to the practical application of satellite technology.
	Oct	Policy of the four modernisations and rectification; restoration of the chief designer system.
		Start of construction of new launch site, Xi Chang.
1980	May 18	First operational test of the Dong Feng 5 intercontinental ballistic missile, impacting in the South Pacific (second test, 21 May).
		China joins International Astronautical Federation.
1981	Sep 20	China launches three scientific satellites, Shi Jian 2, 2A and 2B; the third country after the Soviet Union and the United States to launch three satellites on one rocket.
1982	Apr 9	Establishment of the Space Ministry, replacing the Seventh Ministry
	Oct 12	First firing of submarine launched intercontinental ballistic missile.
1984	Jan 29	First launch of the Long March 3; China becomes, after the United States and Europe, the third country to master liquid hydrogen fuels.; Dong Fang Hong 2 communications satellite put, incorrectly, in low Earth orbit; first launch from new Xi Chang launch site.
	Apr 8	China successfully puts a 24-hour communications satellite into orbit, the Shiyan Tongbu Tongxin Weixing (experimental geostationary communications satellite).

Year	Date	Event
1985	Oct	China puts its rockets on the world launcher market.
1986	Feb 1	First operational geostationary communications satellite Shiyong Tongbu Tongxin Weixing.
		Start of construction of Haikou launch site on Hunan Is, for sounding rockets.
		Building of Miyun Earth station to receive Landsat data.
1987	Sep 9	Introduction of FSW 1 series of recoverable satellites – heavier, able to stay in orbit up to 10 days; the series introduced microbiology experiments in addition to remote sensing tasks.
1988	Sep 6	First Chinese meteorological satellite, Feng Yun 1; first launch of the Long March 4; first launch from new Taiyuan launch site.
	Feb 1	First of the Dong Fang Hong 2A communications satellites.
	Dec 19	Launch of Zhinui (Weaver Girl) 1 sounding rocket from Haikou.
1990	Apr 7	First commercial launch by Chinese Long March 3 rocket (Asiasat, for a Hong Kong company).
	Jul 16	First launch of new Long March 2E launcher, putting Pakistan satellite (Badr) into orbit.
	Oct 5	China flies animals into orbit on FSW 1-3, the third country, after the Soviet Union and the United States, to do so.
1991	Jan 22	First launch of Weaver Girl 3 sounding rocket.
1992	Aug 9	Introduction of FSW 2 series of recoverable satellites – heavier, manoeuvrable, able to stay in orbit up to 18 days.
1993		Establishment of Chinese National Space Administration (CNSA).
		Chief of staff of the People's Liberation Army, Chi Haotian, visits Star Town, the cosmonaut training centre in Moscow, marking renewal of contact broken in 1960.
1994	Feb 8	First launch of Long March 3A, carrying Shi Jian 4 scientific satellite.
1995	Mar	Commissioning of comship *Yuan Wang 3.*
1996	Feb 14	Long March 3B rocket crashes on its first flight – the Saint Valentine Day's massacre.
	Mar	FSW 1-5 crashes into the South Atlantic; launched in October 1993, it had failed to return to Earth and had blasted into an unstable orbit.
	Aug 20	Arrival of Chinese cosmonaut trainers and other specialists in Star Town, Moscow.

	Oct	47th International Astronautical Federation Congress held in Beijing. Plans announced for manned spaceflight in period 1999–2001 (project 921).
1997	May 8	First successful launch of the Dong Fang Hong 3 series of communications satellites.
	Jun 10	Launch of first geostationary meteorological satellite, Feng Yun 2.
	Aug 20	First successful flight of Long March 3B.

List of all Chinese launchings

Satellites entering orbit

Chinese name	Date	Weight (kg)	Incl. (°)	Per. (min)	Peri/Apo. (km)	Launcher	Site
1 Dong Fang Hong	24 Apr '70	173	68.5	114	440–2,386	CZ-1	Jiuquan
2 Shi Jian 1	3 Mar '71	221	69.9	106	267–1,830	CZ-1	Jiuquan
3 Ji Shu Shiyan Weixing 1	26 Jul '75	1,107	69.9	91	183–460	FB-1	Jiuquan
4 Fanhui Shi Weixing 0-1	26 Nov '75	1,790	63	91	179–479	CZ-2C	Jiuquan
5 Ji Shu Shiyan Weixing 2	16 Dec '75	1,109	69	90.2	186–387	FB-1	Jiuquan
6 Ji Shu Shiyan Weixing 3	30 Aug '76	1,108	69.1	108	198–2,145	FB-1	Jiuquan
7 Fanhui Shi Weixing 0-2	7 Dec '76	1,790	59.5	91.1	171–480	CZ-2C	Jiuquan
8 Fanhui Shi Weixing 0-3	26 Jan '78	1,810	57	90.9	161–479	CZ-2C	Jiuquan
9 Shi Jian 2	19 Sep '81	257	59.5	103.4	390–1,600	FB-1	Jiuquan
Shi Jian 2A		483		103.5	235–1,615		
Shi Jian 2B		28		103.2	232–1,597		
10 Fanhui Shi Weixing 0-4	9 Sep '82	1,780	62.9	90.2	174–393	CZ-2C	Jiuquan
11 Fanhui Shi Weixing 0-5	19 Aug '93	1,840	63.3	90.1	172–389	CZ-2C	Jiuquan
12 Shiyan Weixing	29 Jan '84	461	31	92	290–460	CZ-3	Xi Chang
13 Shiyan Tongbu Tongxin Weixing	8 Apr '84	461	0.72	1,444	35,521–36,383	CZ-3	XiChang
14 Fanhui Shi Weixing 0-6	12 Sep '84	1,810	67.9	90.3	174–400	CZ-2C	Jiuquan

Satellites entering orbit, Cont/d

Chinese name	Date	Weight (kg)	Incl. (°)	Per. (min)	Peri/Apo. (km)	Launcher	Site
15 Fanhui Shi Weixing 0-7	21 Oct '85	1,810	62.9	90.2	171–393	CZ-2C	Jiuquan
16 Shiyong Tongbu Tongxin Weixing 1	1 Feb '86	433	0.17	1,450	35,895–36,225	CZ-3	Xi Chang
17 Fanhui Shi Weixing 0-8	6 Oct '86	1,770	57	90	173–385	CZ-2C	Jiuquan
18 Fanhui Shi Weixing 0-9	5 Aug '87	1,810	62.9	90.2	172–400	CZ-2C	Jiuquan
19 Fanhui Shi Weixing 1-1	9 Sep '87	2,070	63	89.7	206–310	CZ-2C	Jiuquan
20 Shiyong Tongbu Tongxin Weixing 2	7 Mar '88	441	0.54	1,455	35,784–36,612	CZ-3	Xi Chang
21 Fanhui Shi Weixing 1-2	5 Aug '88	2,130	63	89.7	206–310	CZ-2C	Jiuquan
22 Feng Yun 1-1	6 Sep '88	757	99.1	102.83	881–905	CZ-4A	Taiyuan
23 Shiyong Tongbu Tongxin Weixing	22 Dec '88	441	0.53	1,471	35,756–337,180	CZ-3	Xi Chang
24 Shiyong Tongbu Tongxin Weixing 4	4 Feb '90	441	0.45	1,472	35,780–37,199	CZ-3	Xi Chang
25 Asiasat 1	7 Apr '90	1,250	0.26	1,460	35,791–36,744	CZ-3	Xi Chang
26 Badr	16 Jul '90	52	28.5	96.6	205–983	CZ-2E	Xi Chang
Aussat model		7,353		did not orbit			
27 Feng Yun 1 -2	3 Sep '90	889	98.9	102.8	885–900	CZ-4A	Taiyuan
Qi Qui Weixing 1		8	98.9	102.8	884–900		
Qi Qui Weixing 2		8	98.9	102.6	862–902		
28 Fanhui Shi Weixing 1-3	5 Oct '90	2,080	57	89.7	208–311	CZ-2C	Jiuquan
29 Shiyong Tongbu Tongxin Weixing 5	28 Dec '91	1,024	31.1	112.4	219–2,451	CZ-3	Xi Chang
30 Fanhui Shi Weixing 2-1	9 Aug '92	2,590	63.17	89.5	172–330	CZ-2D	Jiuquan
31 Optus B-1	13 Aug '92	1,582	0.33	1,472	35,657–37,330	CZ-2E	Xi Chang
32 Fanhui Shi Weixing 1-4	6 Oct '92	2,060	63	89.8	214–312	CZ-2C	Jiuquan
Freja		259	63	109	596–1,763		
33 Fanhui Shi Weixing 1-5	8 Oct '93	2,100	57	89.6	209–300	CZ-2C	Jiuquan

Satellites entering orbit, Cont/d

Chinese name	Date	Weight (kg)	Incl. (°)	Per. (min)	Peri/Apo. (km)	Launcher	Site
34 Shi Jian 4	8 Feb '94	400	28.5	638	209–36,118	CZ-3A	Xi Chang
KF-1		1,600	28.5	636	210–36,054		
35 Fanhui Shi Weixing 2-2	3 Jul '94	2,760	63	89.7	174–343	CZ-2D	Jiuquan
36 Apstar 1	21 Jul '94	1,368	0.3	1,579	35,269–41,820	CZ-3	Xi Chang
37 Optus B-3	27 Aug '94	1,700	1.56	1,389	30,778–38,978	CZ-2E	Xi Chang
38 Zhongxing 6	29 Nov '94	2,232	0.26	1,426	35,181–35,993	CZ-3A	Xi Chang
39 Asiasat 2	28 Nov '95	1,400	0.6	1,404	34,375–35,932	CZ-2E	Xi Chang
40 Echostar 1	28 Dec '95	3,288	5	981	17,572–35,072	CZ-2E	Xi Chang
41 Apstar 1A	7 Mar '96	1,400	1	1,553	34,089–42,010	CZ-3	Xi Chang
42 Zhongxing 7	18 Aug '96	1,200	27	307.5	200–17,229	CZ-3	Xi Chang
43 Fanhui Shi Weixing 2-3	20 Oct '96	2,970	63	89.6	171–342	CZ-2D	Jiuquan
44 Dong Fang Hong 3-2	8 May '97	2,232	0.32	1,436	35,778–35,788	CZ-3A	Xi Chang
45 Feng Yun 2	10 Jun '97	1,380	1.19	1,436	35,780–35,792	CZ-3	Xi Chang
46 Agila	20 Aug '97	3,770	0.3	1,436	35,777–35,791	CZ-3B	Xi Chang
47 SD test	1 Sep '97	650 650	87	97.3	189–675	CZ-2C-SD	Taiyuan
48 Apstar 2R	16 Oct 97	3,747	0.1	1442.8	35,818–36,018	CZ-3B	Xi Chang
49 Iridium	8 Dec 97	650 650				CZ-2C-SD	Taiyuan

Launch failures

Date	Launcher	Intended payload	Outcome
14 Jul 1974	Feng Bao 1	Military satellite	Second stage failed.
5 Nov 1974	Long March 2A	Recoverable satellite	First stage failed.
28 Jul 1979	Feng Bao 1	Shi Jian scientific satellites	Second stage failed.
21 Dec 1992	Long March 2E	Optus B-2	Satellite lost during launch.
25 Jan 1995	Long March 2E	Apstar 2	Exploded at 70 s.
14 Feb 1996	Long March 3B	Intelsat 708	Guidance failed at 2 s; exploded.

Glossary

The terminology of the Chinese space programme may not be well known in the West. These glossaries contain terms associated with Chinese satellites, rockets and space equipment, the main institutional bodies and academies associated with the space programme, an introduction to the key personalities of the programme and China's political leadership, and the main political terms and developments of the period which had an impact on the space programme. A brief listing of some technical terms of space terminology is also included.

Chinese satellites, rockets, space equipment and related terms

Chang Zhen	'Long March', the name of China's main family of rocket launchers.
Dong Fang Hong	'The East is Red'. This is the name of China's first satellite. Much later, it was also the name of China's communications satellite. These were called the Dong Fang Hong 2, 2A and 3 series.
Dong Feng	'East wind', the name of China's missiles (Dong Feng 1, 2 etc).
Fanhui Shi Weixing	Recoverable experimental satellite, series beginning in 1975.
Feng Bao	'Storm', the name of China's launcher developed in Shanghai.
He Ping	'Peace', the name of a series of Chinese sounding rockets (He Ping 2 and He Ping 6).
Ji Shu Shiyan Weixing	Technical experimental satellite (series of three satellites, probably military, launched 1975–6)
Jiuquan	China's first launch site, in the Gobi desert.
Qi Qui Weixing	Atmospheric satellite (balloons carried into orbit on Feng Yun 1-2 in 1990).

Shi Jian	Name for China's series of scientific satellites. In Chinese, the word Shi Jian means 'practice' and 'construction'.
Shiyan Weixing	'Experimental satellite' (communications satellite launched 29 January 1984)
Shiyan Tongbu Tongxin Weixing	'Experimental geostationary communications satellite', launched 8 April 1984.
Shiyong Tongbu Tongxin Weixing	'Operational geostationary communications satellite', first flown 1988.
Taiyuan	China's third launch site, used for satellites flying into polar, Sun-synchronous orbit (*qv*).
Xi Chang	China's second launch site, in south-western China, used for geostationary satellites (*qv*).
Yeti fadong	Liquid-fuel rocket engine. Chinese rocket engines are designated YF-1 and so on. 'GF' is the acronym used for solid fuel rocket engines.
Zhinui	'Weaver Girl', the name of a Chinese sounding rocket programme from 1988.
Zhongxing	'The star of China'. Zhongxing 5 was the name given to an American satellite already in orbit but bought by the Chinese in 1992. Zhongxing 6 was an abandoned comsat launched by the Chinese in 1994. Zhongxing 7 was an American satellite bought in 1996 and launched by China, but which entered the wrong orbit and was abandoned. Zhongxing 1 to 4 were never identified and the classification may now be concluded.

Institutions and academies

ARMT	Chinese Academy for Solid Rocket Motors.
CASC	China Aerospace Corporation.
CASET	Chinese Academy for Space Electronics.
CALT	Chinese Academy of Launcher Technology, formerly Beijing Wan Yuang Industrial Corporation.
CAST	Chinese Academy of Space Technology.
CCF	Chinese Academy of Mechanical and Electrical Engineering.
CHETA	Chinese Electromechanic Academy.
CLTC	China Satellite Launch and Tracking General Control.

Fifth Academy — The main organisational body responsible for the Chinese space programme from 1956 to 1964.

Gt Wall Industries Corp — The main company promoting Chinese launchers and commercial space products in the West.

SAST — Shanghai Academy of Space Technology.

Seventh Ministry — The main organisational body responsible for the Chinese space programme from 1964 to 1982.

SIMED — Shanghai Institute of Machine and Electrical Design

Key personalities

Deng Xiaoping — Leader of China from 1978 to 1997. He tried to restore order in China after the death of Mao, opening the economy.

Lin Biao — Leader of a leftist group in the Chinese leadership, supportive of the Cultural Revolution, influential in the Military Commission in 1970. The following year, Lin Biao fled China for the Soviet Union, but his plane was shot down in flames before he reached the border.

Mao Zedong — Leader of the Chinese Revolution (1949). He died in September 1976.

Nie Rongzhen — The military leader of the Chinese space programme.

Tsien Hsue-shen — The father of the Chinese space programme.

Zhou Enlai — Foreign minister, then Prime Minister under Mao Zedong, considered to be a moderating influence during the Cultural Revolution. He died shortly before Mao, in April 1976.

Key political events and terms

Cultural Revolution — This was launched by Chairman Mao Zedong in March 1966. He encouraged young people, who formed the Red Guards, to reinforce the communist revolution by seeking out counter-revolutionary elements and thought. The campaign of radical egalitarianism lasted ten years and was disruptive of the economy and science.

Gang of Four — Four people, led by Mao's wife, Jian Qing, who tried to maintain the revolutionary path of the Cultural Revolution. They seized power after Mao's death but were overthrown by a military coup a month later (October 1976).

Four modernisations — The modernisation of China announced by Deng Xiaoping in 1978. The four modernisations were science, agriculture, industry and defence.

Great leap forward — A national campaign launched by Chairman Mao Zedong in 1958, whereby China would rapidly increase its agriculture and industrial production. Steel was requisitioned from every home, with small furnaces set up in every street. A campaign was announced against pests – mosquitoes, flies, rats and sparrows. Within a year, the country was in chaos, people were starving and the campaign was called off.

Long March — The time in 1932 when Mao Zedong led his communist army 8,000 km out of a nationalist government encirclement to the north of the country, where they subsequently regrouped.

Rectification — The period from 1978 when Deng Xiaoping restored the order to the economy and society that had been upset by the period of the Cultural Revolution.

Third line regions — Inland regions of China, away from the coastal regions and border with the Soviet Union, where industries were dispersed in the 1960s because of the threat of war.

Space terminology

Apogee — The furthest point from Earth in a satellite's orbit.

Geostationary orbit — An orbit 36,000 km high in which a satellite orbits the Earth once every 24 hours above the equator, thus appearing to hover over the same point continuously. It is a perfect altitude for satellites designed for global communications. This is also called a 24-hour orbit.

Inclination — The angle in degrees (°) at which a satellite crosses the equator while orbiting the Earth. This also defines the parts of Earth over which the satellite orbits. Thus, a satellite orbiting at 58° will fly directly over the land and sea mass of Earth between latitude 58°N and 58°S.

Perigee — The closest point to Earth in a satellite's orbit.

Period — The time which a satellite takes to orbit the Earth.

Sun-synchronous orbit — An orbit, generally polar, which passes over the same point at the same time each day, ensuring a constant Sun angle on the target or weather being photographed.

Bibliography (in English)

Books

Irish Chang. *The thread of the silkworm*. Basic Books, New York, 1995.
Heyi Chen. *Into outer space*. China Pictorial Publications, Beijing, 1989.
Phillip S. Clark. *Chinese space activity, 1987–8*. Astro Info Services, London, 1989.
Phillip S. Clark. *China's recoverable satellite programme*. Molniya Space Consultancy, Heston, 1994.
Phillip S. Clark. *Chinese launch vehicles*. Molniya Space Consultancy, Whitton, 1996.
Phillip S. Clark. *The Chinese space programme – an overview*. Molniya Space Consultancy, Whitton, 1996.
Kenneth Gatland. *Missiles and rockets*. Blandford Press, London, 1975.
Kenneth Gatland (*ed.*). *Illustrated Encyclopaedia of Space Technology*. Salamander, London and New York, 1989 (second edition).
Jim Harford and Wilbur Pritchard. *China space report*. New York, American Institute of Aeronautics and Astronautics, 1980.
Jiaqi Song *et al*. *China defence, research and development*. Hong Kong, China Promotion Ltd., Hong Kong, (undated).
Reginald Turnill. *The observer's book of unmanned spaceflight*. Frederick Warne and Co Ltd, London and New York, 1976
Zhu Yilin. *An introduction to Chinese space endeavour* (collected works). China Academy of Space Technology, Beijing, 1995.
Jun Zhang (*ed.*). *The Chinese space industry today*. China Social Sciences Publishing Co, Beijing, 1986 (in four volumes).

Articles, journals and monographs

Gerald L Borrowman: China's long march to orbit. *Spaceflight*, **25**, (5), May 1983.
Yanping Chen. China's space commercialisation effort – organisation, planning and strategies. *Space policy*, 1, February 1993.
Robert Christy. Chinese puzzle no more. *Spaceflight News*, 35, November 1988.
Phillip S. Clark: The Chinese space year, 1984. *Journal of the British Interplanetary Society*, **39**, (1), January 1986.

Phillip S. Clark. China – in business and advancing fast. *Spaceflight*, **29**, (2), February 1987.
Phillip S. Clark. China speeds up in the space race. *Jane's Intelligence Review*, **9**, (4), April 1997.
Sun Jiadong. Lifting off with flying colours. *China in focus*, 21.
Vincent Kohler. China's new long march. *Analog*, 1987.
Hormuz P Mama. China's Long March family of launch vehicles. *Spaceflight*, **37**, (9), September 1995.
G. Lynwood May. China advances in space. *Spaceflight*, **30**, (11), November 1988.
Théo Pirard. Chinese secrets orbiting the Earth. *Spaceflight*, **19**, (10), October 1977.
Lawrence Stern and Jack High: America takes a long march into space. *Spaceflight*, **32**, (4), April 1980.
Zhu Yilin: Space microgravity scientific experiments in China. *Spaceflight*, **35**, (10), October 1993.
Zhu Yilin. Development of Chinese Earth satellites under Prof. Tsien. *Journal of the British Interplanetary Society*, **50**, (5), 1997.
Zhu Yilin. Fast track development of space technology in China. *Space policy*, May 1996.
Zhu Yilin. Applications of remote sensing satellites in China. *Earth space review*. **5**, (3), 1996.
Zhu Yilin with Xu Fuxiang: Status & prospects of China's communication broadcast satellites. *Space policy*. **13**, (1), February 1997.
Zhu Yilin. Development of small satellites and their operation in the Asia-Pacific region. Monograph.
Zhu Yilin. Some results of Chinese space microgravity life science experiments. Monograph.

Periodicals

Air and Cosmos (in French)
Aviation Week and Space Technology
China Pictorial
China Today

Index

WILEY-PRAXIS SERIES IN SPACE SCIENCE AND TECHNOLOGY

Forthcoming Titles

THE SPACE SHUTTLE: Roles, Missions and Accomplishments
David M. Harland, formerly Visiting Professor, University of Strathclyde, UK

THE SPACE DEBRIS ENVIRONMENT: Hazard and Risk Assessment
Nicholas L. Johnson, NASA Johnson Space Center, Houston, Texas, USA, *et al.*

SOLAR SAILINIG: Technology, Dynamics and Mission Applications
Colin R. McInnes, Department of Aerospace Engineering, University of Glasgow

SPACE GOVERNANCE: A Blueprint for Future Activities
George S. Robinson, Attorney-at-Law, President, Ocean-Space Services, Adjunct Professor, George Mason University, Institute of International Transactions, Virginia, USA and Declan O'Donnell, Attorney-at-Law, President of the World Bar Association, President and Founder of the United Societies in Space, USA

THE MOON: RESOURCES, FUTURE DEVELOPMENTS AND COLONIZATION
David G. Schrunk, formerly Radiation Safety Officer and Head of Nuclear Medicine, Polomar-Pomerado Hospital District, Escondido, California, Founder and Chairman, Science of Laws Institute, San Diego, California, USA; Burton L. Sharpe, System Sales Engineer, Communications Corporation, St Louis, MO, formerly Resident Site Engineer, NASA/Jet Propulsion Laboratory, supporiting US Transportation Command, Scott AFB ILUSA; Bonnie L. Cooper, Scientist, Oceaneering Space Systems, Houston, Texas

ROCKET AND SPACECRAFT PROPULSION: Principles, Practice and New Developments
M.J.L. Turner, Principal Research Fellow, Department of Physics and Astronomy, University of Leicester, UK